Lecture Notes in Computer Science 10739

Commenced Publication in 1973
Founding and Former Series Editors:
Gerhard Goos, Juris Hartmanis, and Jan van Leeuwen

More information about this series at http://www.springer.com/series/7407

Dan Alistarh · Alex Delis
George Pallis (Eds.)

Algorithmic Aspects of Cloud Computing

Third International Workshop, ALGOCLOUD 2017
Vienna, Austria, September 5, 2017
Revised Selected Papers

 Springer

Editors
Dan Alistarh
IST AUSTRIA
Klosterneuburg
Austria

Alex Delis
University of Athens
Athens
Greece

George Pallis
University of Cyprus
Nicosia
Greece

ISSN 0302-9743   ISSN 1611-3349 (electronic)
Lecture Notes in Computer Science
ISBN 978-3-319-74874-0   ISBN 978-3-319-74875-7 (eBook)
https://doi.org/10.1007/978-3-319-74875-7

Library of Congress Control Number: 2018931375

LNCS Sublibrary: SL1 – Theoretical Computer Science and General Issues

Printed on acid-free paper

This Springer imprint is published by the registered company Springer International Publishing AG
part of Springer Nature
The registered company address is: Gewerbestrasse 11, 6330 Cham, Switzerland

Preface

The International Workshop on Algorithmic Aspects of Cloud Computing (ALGOCLOUD) is an annual event aiming to tackle the diverse new topics in the emerging area of algorithmic aspects of computing and data management in the cloud.

The aim of the workshop is to bring together international researchers, students, and practitioners to present research activities and results on topics related to algorithmic, design, and development aspects of modern cloud-based systems.

As in prior years, paper submissions were solicited through an open call for papers. ALGOCLOUD welcomes submissions on all theoretical, design, and implementation aspects of modern cloud-based systems. We are particularly interested in novel algorithms in the context of cloud computing, cloud architectures, as well as experimental work that evaluates contemporary cloud approaches and pertinent applications. We also welcome demonstration manuscripts, which discuss successful elastic system developments, as well as experience/use-case articles. Contributions may span a wide range of algorithms for modeling, practices for constructing, and techniques for evaluating operations and services in a variety of systems, including but not limited to virtualized infrastructures, cloud platforms, datacenters, cloud-storage options, cloud data management, non-traditional key-value stores on the cloud, HPC architectures, etc.

Topics of interest addressed by this workshop include, but are not limited to:

- Algorithmic aspects of elasticity
- Search and retrieval algorithms for cloud infrastructures
- Scale-up and -out for NoSQL and columnar databases
- Resource provisioning and management
- Monitoring and analysis of elasticity for virtualized environments
- Analysis of containerized applications
- Cloud deployment tools and their analysis
- Query languages and novel programming models
- Content delivery through cloud infrastructures
- Load-sharing and caching for cloud systems
- Data structures and algorithms for eventually consistent stores
- Scalable access structures and indexing for cloud data stores
- Algorithmic aspects for cloud applications
- Machine learning, analytics, and data science
- Resource availability, reliability, and fail-over
- NoSQL and schema-less data modeling and integration
- Consistency, replication, and partitioning CAP
- Transactional models and algorithms for cloud data stores

ALGOCLOUD 2017 took place on September 5, 2017, in Vienna, Austria. It was collocated and was a part of ALGO 2017 (September 4–8, 2017), the major annual congress that combines the premier algorithmic conference European Symposium on Algorithms (ESA) and a number of other specialized symposiums and workshops and a summer school, all related to algorithms and their applications, making ALGO the major European event for researchers, students, and practitioners in algorithms.

The Program Committee (PC) of ALGOCLOUD 2017 was delighted by the positive response to the call for papers. The diverse nature of papers submitted, demonstrated the vitality of the algorithmic aspects of cloud computing. All submissions underwent the standard peer-review process and were reviewed by at least four PC members. The PC decided to accept nine original research papers in a wide variety of topics that were presented at the workshop. We would like to thank the PC members for their significant contribution in the reviewing process.

The program of ALGOCLOUD 2017 was complemented with a highly interesting keynote, entitled "The Clouds Have Taken Over, but Algorithms Are Here to Save the Day," which was delivered by Babak Falsafi (EPFL, Switzerland). We wish to express our sincere gratitude to this distinguished professor for the excellent keynote he provided.

We would like to thank all authors who submitted their research work to ALGOCLOUD. We also thank the Steering Committee for volunteering their time.

We hope that these proceedings will help researchers, students, and practitioners to understand and be aware of state-of-the-art algorithmic aspects of cloud computing, and that they will stimulate further research in the domain of algorithmic approaches in cloud computing in general.

September 2017

Dan Alistarh
Alex Delis
George Pallis

Organization

Steering Committee

Spyros Sioutas	Ionian University, Greece
Peter Triantafillou	University of Glasgow, UK
Christos D. Zaroliagis	University of Patras, Greece

Workshop Chairs

Dan Alistarh	IST, Austria
Alex Delis	University of Athens, Greece

Proceedings and Publicity Chair

George Pallis	University of Cyprus, Cyprus

Program Committee

Stergios Anastasiadis	University of Ioannina, Greece
Athman Bouguettaya	University of Sydney, Australia
Marco Canini	KAUST, Saudi Arabia
Aleksandar Dragojevic	Microsoft Research, UK
Schahram Dustdar	TUW, Austria
Rachid Guerraoui	EPFL, Switzerland
Gabriel Istrate	West University of Timisoara and the e-Austria Research Institute, Romania
Thomas Karagiannis	Microsoft Research, UK
Nectarios Koziris	NTUA, Greece
Fernando Pedone	University of Lugano, Switzerland
Florin Pop	University Politehnica of Bucharest, Romania
Raj Ranjan	Newcastle University, UK
Luis Rodrigues	Universidade Técnica de Lisboa, Portugal
Rizos Sakellariou	University of Manchester, UK
Stefan Schmid	Aalborg University, Denmark
Zahir Tari	RMIT, Australia
Vasileios Trigonakis	Oracle Labs, Switzerland
Dimitris Tsoumakos	Ionian University, Greece

Contents

X   Contents

Invited Paper

Warehouse-Scale Computing in the Post-Moore Era

Babak Falsafi$^{(\boxtimes)}$

EcoCloud and EPFL, Lausanne, Switzerland
babak.falsafi@epfl.ch

Todays IT services are provided by centralized infrastructure referred to as datacenters. In contrast to supercomputers aimed at the high-cost/high-performance scientific domain, datacenters consist of low-cost servers for high-volume data processing, communication and storage. Datacenter owners prioritize capital and operating costs (often measured in performance per watt) over ultimate performance.

Modern full-size datacenters are warehouse-scale facilities that consume up to 30 MW of electricity, are more than 20 times the size of a football field and cost several billions of dollars. These facilities are at physical limits defined by sources of electricity, networking and cooling. While servers currently comprise less than 2% of worldwide electricity consumption, their share of electricity in modern information-based economies is higher and while datacenter efficiencies are improving over the years the combined impact of growth and the overall life-cycle energy (including manufacturing) is growing fast.

The emergence of massive datacenters is a direct response to an IT revolution, at the center of which is data. As individuals, we need ubiquitous access, exchange and sharing of data with those we interact with. Similarly, businesses, governments and societies rely on collecting, analyzing and exchanging data to improve their products, services and ultimately enhance our lives. Data now also lies at the core of the supply-chain for both products and services in modern economies. Data-centric science, through analytics on massive data sets, now complements theoretical, empirical and simulation-driven science as a fourth paradigm for scientific discovery. The net result is that our daily demands to store, communicate, and analyze data are growing at much higher rates than improvements in semiconductor technology can sustain.

In 2014, Amazon Web Services added enough server capacity every day as was needed for its global operation in 2004, when it was $7 billion in annual revenue. IDC [1] projects that in the decade ending in 2015, the world generated around 8 ZB of data, a 100x increase over 2005. This observation demonstrates data is growing more rapidly than conventional silicon density, which doubles every two years according to Moore's Law.

This unprecedented growth in the demand for servers and datacenters is on a direct collision course with the slowdown in silicon scaling, the very phenomenon that has led to the proliferation of IT in the past five decades. Since the inception of silicon-based integrated circuits, digital platforms have enjoyed a doubling in density in every two years (Moore's Law) and performance every 18 months with

© Springer International Publishing AG, part of Springer Nature 2018
D. Alistarh et al. (Eds.): ALGOCLOUD 2017, LNCS 10739, pp. 3–7, 2018.
https://doi.org/10.1007/978-3-319-74875-7_1

a roughly constant power envelope. Vendors managed to maintain a constant power envelope through lower transistor supply (and commensurately threshold) voltages, which has been termed Dennard's Law. Recent years have seen a drastic slowdown in voltage scaling due to quantum effects as the transistor channel has reached dimensions on the order of tens of nanometers, resulting in a dramatic shift towards platform designs that are more energy-efficient [8].

Similarly, although density scaling has continued, it has also slowed, and conventional scaling in logic density and memory capacity will soon come to a halt [9]. If not mitigated, the slowdown in silicon scaling will severely limit our ability to tap into the nearly limitless opportunities that massive data can offer.

Therefore, a complete rethinking of server architecture is needed to overcome density and efficiency challenges in datacenters with implications on all layers of the computation stack including algorithms. We envision platform scalability in the post-Moore era along three axes: (1) Integration, technologies for tighter integration of components (from algorithms to infrastructure) to minimize overhead in moving data, (2) Specialization, technologies to accelerate critical services, map algorithms into representations that are closer to the platform and avoid overhead in general-purpose processing and abstractions, and (3) Approximation, technologies to gauge and tune computational output quality to optimize resources used in computation. I refer to these as the ISA approach to post-Moore platform design.

1   Heterogeneity in Platforms

Due to the abundant parallelism present in server workloads, the emergence of manycore chips has, for some time, appeared to make servers immune to scalability concerns. As technology affords more cores on chip, servers were believed to be capable of trivially scaling to the parallelism made available by the hardware and expanding to meet growing IT demands.

Figure 1 shows the number of general-purpose high-performance cores (GPP) and simple embedded (EMB) cores that can be integrated on a die over multiple technology generations based on the available power budget in servers [4,6]. Current server processor parts are already dark or dim because they dedicate up to half of the on-chip area to power-efficient caches instead of cores; unfortunately, such large caches are ineffective at improving the performance of a wide spectrum of server workloads [5]. Therefore, despite the inherent scalability in multithreaded server workloads, increasing core counts cannot directly translate into performance improvements, because the system software stack does not scale up and chips are physically constrained in power and off-chip bandwidth. Meanwhile, novel approaches to silicon density scaling will continue albeit at an extremely slow rate thanks to 3D transistor geometries and die-stacking.

Moreover, unprecedented investment in silicon technologies will keep CMOS around for at least a few decades before novel materials become cost-effective enough to present an alternative to silicon. The slowdown in efficiency, however, will result in platforms that are not only parallel but are heterogeneous with

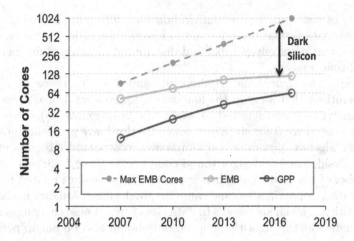

Fig. 1. Power envelope leading to Dark Silicon [DarkServers].

a spectrum of accelerators and application programming interfaces emerging not only on all digital platforms but in particular in servers where historically software and system stacks are deep with well-defined abstractions and interfaces at every level of the stack.

Similarly, novel memory and storage technologies are emerging that present a scalability path for conventional memory but have distinct enough traits that cause a potential disruption in platform design. Emerging memory allows for nearby logic to offload data-intensive tasks but appears to logic as a block-oriented device (better suited for streaming than random access). Emerging memory is also asymmetric in read/write performance and power thereby favoring paradigms that are read intensive. Finally emerging memory is non-volatile impacting persistent data management and requiring careful attention to ordering accesses. These technologies will also introduce heterogeneity for memory and storage leading to both opportunities and challenges in the post-Moore era. These trends apply to all market segments for digital platforms and reinforce the emergence and convergence of volume servers in warehouse-scale computers as the building block for high-performance computing platforms.

2 Massive Data Analytics

Sophisticated analytic tools beyond indexing and rudimentary statistics (e.g., relational and semantic interpretation of underlying phenomena) over the emerging vast repositories of data will not only serve as future frontiers for knowledge discovery in the commercial world but also form a pillar for scientific discovery [7]. The latter is an area where commercial and scientific applications naturally overlap, and high-performance computing for scientific discovery will highly benefit from the momentum in e-commerce. There are a myriad of challenges facing massive data analytics including management of highly distributed data sources, and tracking of data provenance, data validation, mitigating sampling bias and heterogeneity, data format diversity and integrity, integration, security, sharing, visualization, and massively parallel and distributed algorithms for incremental and/or real-time analysis.

With respect to algorithmic requirements and diversity, there are a number of basic operations that serve as the foundation for computational tasks in massive data analytics (often referred to as dwarfs [2] or giants [3]). They include but are not limited to: basic statistics, generalized n-body problems, graph analytics, linear algebra, generalized optimization, computing integrals and data alignment. Besides classical algorithmic complexity, these basic operations all face a number of key challenges when applied to massive data related to streaming data models, approximation and sampling, high-dimensionality in data, skew in data partitioning, and sparseness in data structures. These challenges not only must be handled at the algorithmic level, but should also be put in perspective given projections for the advancement in processing, communication and storage technologies in platforms.

Many important emerging classes of massive data analytics also have real-time requirements. In the banking/financial markets, systems process large amounts of real-time stock information in order to detect time-dependent patterns, automatically triggering operations in a very specific and tight timeframe when some pre-defined patterns occur. Automated algorithmic trading programs now buy and sell millions of dollars of shares time-sliced into orders separated by 1 ms. Reducing the latency by 1 ms can be worth up to $100 million a year to a leading trading house. The aim is to cut microseconds off the latency in which these systems can reach to momentary variations in share prices.

References

1. International data corporation. www.idc.com
2. Asanovic, K., Bodik, R., Catanzaro, B.C., Gebis, J.J., Husbands, P., Keutzer, K., Patterson, D.A., Plishker, W.L., Shalf, J., Williams, S.W., et al.: The landscape of parallel computing research: a view from berkeley. Technical report, Technical Report UCB/EECS-2006-183, EECS Department, University of California, Berkeley (2006)
3. National Research Council, et al.: Frontiers in massive data analysis. National Academies Press (2013)
4. Esmaeilzadeh, H., Blem, E., St. Amant, R., Sankaralingam, K., Burger, D.: Dark silicon and the end of multicore scaling. ACM SIGARCH Comput. Architect. News **39**, 365–376 (2011)
5. Ferdman, M., Adileh, A., Kocberber, O., Volos, S., Alisafaee, M., Jevdjic, D., Kaynak, C., Popescu, A.D., Ailamaki, A., Falsafi, B.: Clearing the clouds: a study of emerging scale-out workloads on modern hardware. SIGARCH Comput. Archit. News **40**(1), 37–48 (2012)
6. Hardavellas, N., Ferdman, M., Falsafi, B., Ailamaki, A.: Toward dark silicon in servers. IEEE Micro **31**(4), 6–15 (2011)
7. Hey, T., Tansley, S., Tolle, K.M., et al.: The fourth paradigm: data-intensive scientific discovery, vol. 1. Microsoft Research Redmond, WA (2009)

8. Horowitz, M.: Scaling, power and the future of CMOS. In: Proceedings of the 20th International Conference on VLSI Design Held Jointly with 6th International Conference: Embedded Systems, VLSID 2007, p. 23. IEEE Computer Society, Washington, DC (2007)
9. Mann, C.C.: The end of moore's law. Technol. Rev. **103**(3), 42–48 (2000)

Optimization for Cloud Services

A Walk in the Clouds: Routing Through VNFs on Bidirected Networks

Klaus-Tycho Foerster(iD), Mahmoud Parham(✉)(iD), and Stefan Schmid(iD)

Aalborg University, Aalborg, Denmark
{ktfoerster,mahmoud,schmiste}@cs.aau.dk

Abstract. The virtualization of network functions enables innovative new network services which can be deployed quickly and at low cost on (distributed) cloud computing infrastructure. This paper initiates the algorithmic study of the fundamental underlying problem of how to efficiently route traffic through a given set of Virtualized Network Functions (VNFs). We are given a link-capacitated network $G = (V, E)$, a source-destination pair $(s, t) \in V^2$ and a set of waypoints $\mathscr{W} \subset V$ (the VNFs). In particular, we consider the practically relevant but rarely studied case of bidirected networks. The objective is to find a (shortest) route from s to t such that all waypoints are visited. We show that the problem features interesting connections to classic combinatorial problems, present different algorithms, and derive hardness results.

1 Introduction

After revamping the server business, the virtualization paradigm has reached the shores of communication networks. Computer networks have broken with the "end-to-end principle" [37] a long time ago, and today, intermediate nodes called *middleboxes* serve as proxies, caches, wide-area network optimizers, network address translators, firewalls, etc. The number of middleboxes is estimated to be in the order of the number of routers [21].

The virtualization of such middleboxes is attractive not only for cost reasons, but also for the introduced flexibilities, in terms of fast deployment and innovation: in a modern and virtualized computer network, new functionality can be deployed quickly in virtual machines on cloud computing infrastructure. Moreover, the composition of multiple so-called Virtual Network Functions (VNFs) allows to implement complex network services known as *service chains* [33]: traffic between a source s and a destination t needs to traverse a set of network functions (for performance and/or security purposes).

However, the realization of such service chains poses a fundamental algorithmic problem: In order to minimize the consumed network resources, traffic should be routed along *short* paths, while respecting capacity constraints. The problem is particularly interesting as due to the need to traverse waypoints, the resulting route is not a simple path, but *a walk*.

In this paper, we are particularly interested in VNF routing on *bidirected networks*: while classic graph theoretical problems are typically concerned with

© Springer International Publishing AG, part of Springer Nature 2018
D. Alistarh et al. (Eds.): ALGOCLOUD 2017, LNCS 10739, pp. 11–26, 2018.
https://doi.org/10.1007/978-3-319-74875-7_2

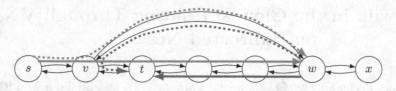

Fig. 1. In this introductory example, the task is to route the flow of traffic from the source s to the destination t via the waypoint w. When routing via the solid red (s, w) path, followed by the solid blue (w, t) path, the combined walk length is $5 + 3 = 8$. A shorter solution exists via the dotted red and blue paths, resulting in a combined walk length of $2 + 2 = 4$. Observe that when the waypoint would be on the node x, no node-disjoint path can route from s to t via the waypoint. Furthermore, some combinations can violate unit capacity constraints, e.g., combining the solid red with the dotted blue path induces a double utilization of the link from v to t. (Color figure online)

undirected or directed graphs, real computer networks usually rely on links providing full-duplex communication. Figure 1 gives an example.

1.1 Model

We study full-duplex networks, modeled as connected *bidirected graphs* $G = (V, E)$ [13] with $|V| = n$ nodes (switches, middleboxes, routers) and $|E| = m$ links, where each link $e \in E$ has a capacity $c : E \to \mathbb{N}_{>0}$ and a weight $\omega : E \to \mathbb{N}_{>0}$. Bidirected graphs (also known as, e.g., Asynchronous Transfer Mode (ATM) networks [10] or symmetric digraphs [22]) are directed graphs with the property that if a link $e = (u, v)$ exists, there is also an anti-parallel link $e' = (v, u)$ with $c(e) = c(e')$ and $\omega(e) = \omega(e')$.

Given (1) a bidirected graph, (2) a source $s \in V$ and a destination $t \in V$, and (3) a set of k waypoints $\mathcal{W} \subset V$, the *bidirected waypoint routing problem* BWRP asks for a flow-route $\mathcal{R}$ (i.e., a walk) from s to t that (*i*) visits all waypoints in $\mathcal{W}$ and (*ii*) respects all link capacities. Without loss of generality, we normalize link capacities to the size of the traffic flow, removing links of insufficient capacity.

BWRP comes in two flavors. In the *unordered* version UBWRP, the waypoints $\mathcal{W}$ can be traversed in any order. In the *ordered* version, OBWRP, the waypoints depend on each other and must be traversed in a pre-determined order: every waypoint w_i may be visited at any time in the walk, and as often as desired (while respecting link capacities), but the route $\mathcal{R}$ must contain a given ordered node sequence $s, w_1, w_2, \ldots, w_k, t$. For example, in a network with stringent dependability requirements, it makes sense first route a packet through a fast firewall before performing a deeper (and more costly) packet inspection.

For both UBWRP and OBWRP, we are interested both in *feasible* solutions (respecting capacity constraints) as well as in *optimal* solutions. In the context of the latter, we aim to optimize the cost $|\mathcal{R}|$ of the route $\mathcal{R}$, i.e., we want to minimize the sum of the weights of all traversed links. As we will see, computing an optimal route $\mathcal{R}^*$ can be hard in general, and hence, we also study α−competitive approximation algorithms, where for any computed tour $\mathcal{R}$ holds: $|\mathcal{R}| \leq \alpha \cdot |\mathcal{R}^*|$.

1.2 Contributions

We initiate the study of the waypoint routing problem on bidirected networks. We put the problem into perspective with respect to classic combinatorial problems (in particular Steiner Tree problems, variants of Traveling Salesman problems, and link-disjoint paths problems), and present a comprehensive set of algorithms and hardness results.

Unordered bidirected waypoint routing UBWRP. We first show that while *any* UBWRP instance is *feasible*, as each link is traversed only once in each direction, computing *optimal* solutions is NP-hard: no polynomial-time approximation scheme (PTAS) exists unless P = NP. On the positive side, by leveraging connections to metric TSP, we show that an ≈1.53-approximation is possible on general graphs. But also *optimal* solutions are possible in polynomial time, namely if the number of waypoints is small, namely $k \in \mathcal{O}(\log n)$: a practically very relevant case. In fact, if the network is planar, we can solve the problem even up to $k \in \mathcal{O}(\log^{2-\varepsilon} n)$ many waypoints in polynomial time (for any constant $\varepsilon > 0$), using a connection to Subset TSP.

Ordered bidirected waypoint routing OBWRP. Due to a connection to link-disjoint paths, it holds that feasible routes can be computed in polynomial time for OBWRP if $k \in \mathcal{O}(1)$. Moreover, while finding optimal routes is NP-hard in general, we show that polynomial-time exact solutions exist for cactus graphs with constant link capacities.

1.3 Related Work

While routing through network functions is a standard application in the area of computer networking, we currently witness the emergence of two new trends which change the requirements on routing: (1) (virtualized) network functions are increasingly deployed not only at the edge but also in the network core [14]; (2) network functions are being composed or "chained" to provide more complex services, a.k.a. *service chains* [19,31,34]. Traversing these network functions may entail certain detours, i.e., routes do not necessarily follow shortest paths and may even contain loops, i.e., form a *walk* rather than a simple path [5,15,16,30].

Waypoint Routing. Amiri et al. [1] recently provided first results on the ordered waypoint routing problem. In contrast to our work, Amiri et al. [1] focus on directed and undirected graphs only and do not consider approximation algorithms; however, finding a feasible solution to the ordered waypoint routing problem is NP-hard on undirected graphs, and for directed graphs already for a single waypoint. The same limitation holds for the work by Amiri et al. [2] on the unordered waypoint problem on undirected graphs. Moreover, under unit capacities, the undirected problem in [2] also has a different structure, and, e.g., it is not possible to establish the connection to Steiner Tree and Traveling Salesman problem variants as discussed in this paper. On the positive side, it is easy to see that their algorithm to compute optimal unordered solutions on bounded treewidth graphs in XP time also applies to our case; however, their hardness results do not.

Link-Disjoint Paths. Closely related to the (ordered) waypoint routing problem is the study of link-disjoint paths: Given k source-destination pairs, is it possible to find k corresponding pairwise link-disjoint paths? For an overview of results on directed and undirected graphs, we refer to the work of Amiri et al. [1].

For bidirected graphs, deciding the feasibility of link-disjoint paths is NP-hard [10]. When the number of link-disjoint paths is bounded by a constant, feasible solutions can be computed in polynomial time [22]. Extensions to multi-commodity flows have been studied in [23,39] and parallel algorithms were presented in [26,27]. To the best of our knowledge, the joint optimization (i.e., shortest total length) of a constant number of link-disjoint paths is still an open problem. In comparison, we can solve UBWRP for a super-constant number of waypoints optimally.

Besides simple graph classes such as trees, we are not aware of algorithms for a super-constant number of link-disjoint paths on directed graphs, which are also applicable to bidirected graphs. We refer to [32] for an annotated tableau. In contrast, we in this paper optimally solve OBWRP on cactus graphs with constant capacity.

Traveling Salesman and Steiner Tree problems. The unordered problem version is related to the *Traveling Salesman problem* (TSP): there, the task is to find a cycle through all nodes of an undirected graph, which does not permit any constant approximation ratio [12], unless P = NP: for the so-called metric version, where nodes may be visited more than once (identical to the TSP with triangle inequality), performing a DFS-tour on a minimum spanning tree (MST) gives a 2-approximation, see, e.g., [12] again. The best known approximation ratio for the metric TSP is 1.5 [11] ($3/2 + 1/34 \approx 1.53$ for $s \neq t$, called the $s - t$ *Path TSP* [38]) and no better polynomial-time approximation ratio than $123/122 \approx 1.008$ is possible [24], unless P = NP. Throughout this paper, when referring to (any variant of) TSP, we usually refer to the metric version on undirected graphs.

A related problem is the NP-hard [18] *Steiner Tree problem* (ST) on undirected graphs: given a set of terminals, construct a minimum weight tree that contains all terminals. If all nodes are terminals, this reduces to the (polynomial) MST problem. The currently best known approximation ratios for ST are $\ln 4 + \varepsilon < 1.39$ [9] (randomized) and $1 + \frac{\ln 3}{2} \approx 1.55$ [36] (deterministic).

Note that both the Steiner Tree and the Traveling Salesman problem are *oblivious* to link capacities. Notwithstanding, we can make use of approximation algorithms for both problems for UBWRP in later sections.

Prize-Collecting and Subset Variants. In particular, if $\mathcal{W} \subsetneq V$, we can utilize the *prize-collecting* versions of (path) TSP and ST, called (PC-PATH2) PCTSP and PCST [3]. Here, every node is assigned a non-negative prize (penalty), s.t. if the node is not included in the tour/tree, its prize is added to the cost. For all three prize-collecting variants above, Archer et al. [3]

provide approximation ratios smaller than 2: (1) PCTSP: $97/49 < 1.979592$, (2) PC-PATH2: $241/121 < 1.991736$, (3) PCST: 1.9672 (randomized) and 1.9839 (deterministic).

Contained in PC-TSP is *Subset TSP*, first proposed in [4, Sect. 6], which asks for a shortest tour through specified nodes [28]. Klein and Marx [29] give a $\left(2^{\mathcal{O}(\sqrt{k}\log k)} + W\right) \cdot n^{\mathcal{O}(1)}$ for planar graphs, where W is the largest link weight. They also point out that classic dynamic programming techniques [6,20] have a runtime of $2^k \cdot n^{\mathcal{O}(1)}$ for general graphs. The path version is not considered, we show how to apply any optimal Subset TSP algorithm to UBWRP with $s \neq t$.

Conceptually related (to the non-metric TSP) is the K-cycle problem, which asks for a shortest cycle through K specified nodes or links, i.e., a vertex-disjoint tour. Björklund et al. [7] give a randomized algorithm with a runtime of $2^K \cdot n^{\mathcal{O}(1)}$.

1.4 Paper Organization

We begin with studying UBWRP in Sect. 2: after giving an introduction to waypoint routing in Sect. 2.1, we give hardness and (approximation) algorithm results via metric and subset TSP in Sect. 2.2. We then consider the ordered case in Sect. 3, tackling constant k in Sect. 3.1, we study hardness in Sect. 3.2, and finally provide a cactus graph algorithm in Sect. 3.3. Lastly, we conclude in Sect. 4 with a short summary and outlook.

2 The Unordered BWRP

We start our study of bidirected waypoint routing with a short *tour d' horizon* of the problem in Sect. 2.1, discussing the case of few waypoints and first general approximations via the Steiner Tree problem. We then follow-up in Sect. 2.2 by showing that UBWRP and metric TSP are polynomially equivalent, i.e., algorithmic and hardness results can be reduced. We also establish connections to the subset and prize-collecting TSP variants.

2.1 An Introduction to (Unordered) Waypoint Routing

First, we examine the case of a *single waypoint* w, which requires finding a shortest $s - t$ route through this waypoint. Note that for a single waypoint, the two problem variants UBWRP and OBWRP are equivalent.

One waypoint: greedy is optimal. We observe that the case of a single waypoint is easy: simply taking two *shortest paths* (SPs) $P_1 = SP(s, w)$ and $P_2 = SP(w, t)$ in a greedy fashion is sufficient, i.e., the route $\mathscr{R} = P_1 P_2$ is always feasible (and thus, also always optimal in regards to total weight).

Suppose this is not the case, that is, $P_1 \cap P_2 \neq \emptyset$. Among all nodes in $P_1 \cap P_2$, let u and v be, resp., the first and the last nodes w.r.t. to the order of visits in $\mathscr{R}$. Let P_i^{xy} denote the sub-path connecting x to y in P_i. Thereby we have $\mathscr{R} = P_1 P_2 = P_1^{su} P_1^{uv} P_1^{vw} P_2^{wu} P_2^{uv} P_2^{vt}$ (Fig. 2). Let $\bar{P}$ be the reverse of any walk

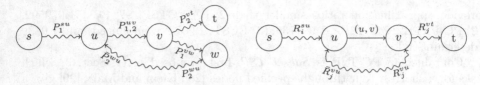

Fig. 2. The directed path from u to v is traversed two times in $\mathscr{R}$.

Fig. 3. The link (u, v) is traversed multiple times in $\mathscr{R}$.

$\mathcal{P}$ obtained by replacing each link $(x, y) \in \mathcal{P}$ with its anti-parallel link (y, x). Observe that for $P_1' = P_1^{su}\bar{P}_2^{wu}$ and $P_2' = \bar{P}_1^{vw}P_2^{vt}$, the feasible route $\mathscr{R}' = P_1'P_2'$ is shorter than $\mathscr{R}$, contradicting P_1 and P_2 both being shortest paths.

Being greedy on the right order is optimal. Next, we show that even more waypoints can be handled efficiently as long as their number is limited: if the optimal traversal order is known, selecting shortest paths in a greedy fashion is again optimal, but to find the optimal order, $k!$ has to be tractable.

We first give an auxiliary lemma, which will also be of later use. We note that Klein and Marx gave an analogous construction for undirected graphs in [29, Fig. 1]. The idea can be explained with Fig. 3: when a link (u, v) is traversed at least twice, we can take a shortcut from u to v. Iterating this idea ensures at most one traversal per link.

Lemma 1. *Any (unfeasible) route $\mathscr{R}$ that traverses some link (u, v) more than once, can be efficiently transformed into a shorter route $\hat{\mathscr{R}}$ that traverses each link at most once.*

Proof. There is at least one link that is being shared by at least two sub-routes along $\mathscr{R}$. For every $k, l : k < l$, denote the set of all links $e \in R_k \cap R_l$ by W. Let $(u, v) \in W$ be a link traversed by R_i and R_j s.t. $i < j = \arg\min_l ((u, v) \in R_i \cap R_l)$ (Fig. 3). Thus, $\mathscr{R} = R_iR_j = R_i^{su}(uv)R_j^{vu}(uv)R_j^{vt}$. Now consider the new route $\mathscr{R}' = R_1^{su}\bar{R}_2^{vu}R_j^{vt}$. Note that $\mathscr{R}'$ might still traverse (u, v) more than once. Repeat the same procedure for (u, v) until this is not the case. Each time we choose the two sub-routes that proceed any other sub-route traversing (u, v). This ensures the other parts of $\mathscr{R}$ will not be skipped after this transformation. Now we remove this link from W. Since we included additional links for the rerouting, there might be some link $(u', v') \in \bar{R}^{vu} \cap \mathscr{R}, (u', v') \neq (u, v)$, that is being traversed more than once in $\mathscr{R}'$, but not in $\mathscr{R}$. Add all such links to W. Repeating the same procedure for every $(x, y) \in W$ transforms $\mathscr{R}'$ into a new route $\mathscr{R}''$ that traverses (x, y) at most once. On the other hand, for each newly included link we exclude its anti-parallel link during the transformation. Therefore the new route is shorter by at least one link, i.e. $|\mathscr{R}''| < |\mathscr{R}|$. Hence, there can be only $O(|\mathscr{R}|)$ iterations, the last of which necessarily ends up at the desired route $\hat{\mathscr{R}}$ satisfying the lemma. □

Using this idea, we can now show that in an optimal ordering, the shortest paths will not overlap.

Lemma 2. *Given the permutation σ of waypoints in the (first visit) order of an optimal route $\mathscr{R}^*$, we can efficiently construct a route $\mathscr{R}_\sigma$ s.t. $|\mathscr{R}_\sigma| = |\mathscr{R}^*|$.*

Proof. Let $\mathcal{W}_\sigma = w_{\sigma_1} w_{\sigma_2} \ldots w_{\sigma_k}$ be the order we intend to visit the waypoints, and $\mathscr{R}'$ be an empty route. For each consecutive pair w_{σ_i} and $w_{\sigma_{i+1}}$ add the shortest path links of $SP(w_{\sigma_i}, w_{\sigma_{i+1}})$ and $SP(w_{\sigma_{i+1}}, w_{\sigma_i})$ to $\mathscr{R}'$. We claim that $\mathscr{R}'$ is a feasible, and constitutes an optimal solution to UBWRP. For the sake of contradiction assume this is not the case. Using Lemma 1 we can construct a feasible route $\hat{\mathscr{R}}$ s.t. $|\hat{\mathscr{R}}| < |\mathscr{R}'|$. This implies that for some shortest path $SP(w_{\sigma_i}, w_{\sigma_j}), |i - j| = 1$, when replaced by the path $\hat{\mathscr{R}}^{w_{\sigma_i} w_{\sigma_j}}$, yields a shorter route, contradicting the optimality of $SP(.)$.   □

This directly implies the following Theorem 3, which is essentially a brute-force approach. We note that implications from the connections between UBWRP and metric TSP in the next section will improve Theorem 3 in such a way that $k \in \mathcal{O}(\log n)$ becomes tractable.

If the order of visits in $\mathscr{R}^*$ was part of the input, then by Lemma 2 the union of shortest paths between consecutive waypoints would be necessarily a feasible and therefore an optimal solution. Since this is not the case (i.e. we cannot know the optimal order in advance), one can iterate over all permutations of waypoints and apply Lemma 2. Then the best of all these iterations will give an optimal route. Thus, for constants c and $c' = c + c \log c$, there are $k! = \left(\frac{c \log n}{\log \log n} \right)! < (c \log n)^{\left(\frac{c \log n}{\log \log n} \right)} = ((c \log n)^{\log n})^{\frac{c}{\log \log n}} < (n^{c \log c} n^c) \in \mathcal{O}(n^{c'})$ iterations. After multiplying by the cost of $O(k)$ calls to $SP(.)$, the polynomial time follows immediately. Thus we have the following theorem:

Theorem 3. *For $k \in \mathcal{O}\left(\frac{\log n}{\log \log n} \right)$, UBWRP is polynomially solvable.*

We now turn our attention to the general case of $k \in \mathcal{O}(n)$. We first establish a connection to (Steiner) tree problems as well as to the Traveling Salesman problem. Subsequently, we will derive stronger results leveraging the metric TSP (Sect. 2.2).

UBWRP is always feasible. Interestingly, UBWRP is always feasible. We begin with the case of $s = t$: first, compute a minimum spanning tree T_U in the undirected version of G, in G_U. Then, traverse T with a DFS-tour starting at s, using every directed link in the bidirected version of T_U exactly once. As every node $v \in V$ will be visited, so will all waypoints $\mathcal{W}$.

The case of $s \neq t$ is similar: Removing the links from the unique $s - t$-path $s, v_1, v_2, \ldots, v_p, t$ decomposes T_U into a forest, still containing all nodes in V. We now traverse as follows: Take a DFS-tour of the tree attached to s, then move to v_1, take a DFS-tour of the tree attached to v_2, …, until lastly arriving at t, then taking a DFS-tour of the tree attached to t, finishing at t.

Approximations via the Steiner Tree problem. For $\mathcal{W} = V$, the above MST approach directly yields a 2-approximation, cf. TSP in [12]. The 2-approximation fails though when $\mathcal{W} \neq V$: e.g., if only one node is a waypoint, visiting all other nodes can add arbitrarily high costs. However, there is a direct duality to the Steiner Tree problem: When setting all waypoints (including s, t) as terminals, an optimal Steiner Tree for these terminals in G_U is a lower bound for an optimal solution to UBWRP: taking the link-set of any route $\mathscr{R}$ in G_U contains the links of a Steiner Tree as a subset. Hence, the construction is analogous to the MST one for $\mathscr{W} = V$. As the best known approximation ratio for the Steiner Tree problem are $\ln 4 + \varepsilon < 1.39$ [9] (randomized) and $1 + \frac{\ln 3}{2} \approx 1.55$ [36] (deterministic), we obtain approximation ratios of $2 \ln 4 + \varepsilon < 2.78$ (rand.) and $2 + \ln 3 \approx 3.09$ (det.), for any constant $\varepsilon > 0$.

2.2   Hardness and Improved Approximation

Next, we show that metric TSP (denoted ΔTSP for the remainder of this section for clarity) is equivalent to our problem of UBWRP on general graphs, in the sense that their corresponding optimal solutions have identical cost.

Theorem 4. *Let I be an instance of* UBWRP *on a bidirected graph G and let I' be an instance of the (path) ΔTSP on the metric closure of the corresponding G_U restricted to $\mathscr{W} \cup \{s, t\}$. The cost of optimal solutions for I and I' are identical.*

Proof. We start with $s = t$ for UBWRP. By setting $V = \mathscr{W} \cup \{s, t\}$, the first reduction follows directly. For the other direction, we first construct an instance of ΔTSP. Then, an optimal solution to ΔTSP must imply an optimal solution to UBWRP and vice versa. Let G'_U be the metric closure of G_U restricted to nodes in $\mathscr{W} \cup \{s, t\}$. An optimal TSP cycle C^* in G'_U (after replacing back the shortest path links) corresponds to a route $\mathscr{R}_{C^*}$ on G_U s.t. $|\mathscr{R}_{C^*}| = |C^*|$. Furthermore, $\mathscr{R}_{C^*}$ possibly violates some link capacities in G. Using Lemma 1 we turn $\mathscr{R}_{C^*}$ to a route $\mathscr{R}'_{C^*}$ feasible for UBWRP. We claim that $\mathscr{R}'_{C^*}$ is optimal. Assume this is not the case, i.e. $|\mathscr{R}'_{C^*}| > |\mathscr{R}^*|$. Let σ be the permutation corresponding to the order of waypoints in $\mathscr{R}^*$. By Lemma 2, we can construct a feasible route $\mathscr{R}_\sigma$, such that $|\mathscr{R}_\sigma| = |\mathscr{R}^*|$ and it uses only the shortest path links chosen by $SP(w_{\sigma_i}, w_{\sigma_{(i+1)} \bmod k}), 0 \le i \le k$. That is, for the cycle C_σ induced by σ on the links of G', we have $|C_\sigma| = \sum_{i=0}^{k} |SP(w_{\sigma_i}, w_{\sigma_{(i+1)} \bmod k})| = |\mathscr{R}_\sigma| \overset{\text{Lemma 2}}{=} |\mathscr{R}^*| < |\mathscr{R}'_{C^*}| \le |\mathscr{R}_{C^*}| = |C^*|$ (by Lemma 1), which contradicts C^* being optimal.

It remains to show, given $\mathscr{R}^*$ and the order of waypoints therein, σ, the cycle C_σ is optimal for ΔTSP (i.e. $|C_\sigma| = |C^*|$). Assume $|C_\sigma| > |C^*|$. As it was shown previously, we can construct a route $\mathscr{R}'_{C^*}$ s.t. $|\mathscr{R}'_{C^*}| \overset{\text{Lemma 1}}{=} |C^*| < |C_\sigma| \overset{\text{Lemma 2}}{=} |\mathscr{R}^*|$, which contradicts the optimality of $\mathscr{R}^*$. The proof construction for $s \ne t$ is analogous, replacing the TSP cycle with a path from s to t. $\square$

No PTAS for UBWRP, but good approximation ratios. As already seen in the proof of Theorem 4, solutions between the corresponding instances of UBWRP and (path) ΔTSP can be efficiently transformed to one of less or identical cost. As such, we can make use of known algorithms and complexity results, resulting in the following two corollaries regarding hardness and approximability:

Corollary 5. UBWRP *is an NP-hard problem, no better polynomial-time approximation ratio than* $123/122 \approx 1.008$ *is possible [24], unless* $P = NP$.

Corollary 6. *For* $s = t$, UBWRP *can be approximated in polynomial time with a ratio of 1.5 [11]. For* $s \neq t$, *a ratio of* $3/2 + 1/34 \approx 1.53$ *can be obtained [38].*

Relations to $(0, \infty)$**-PC-TSP and Subset TSP.** In the prize-collecting (PC) variant, the classical Steiner Tree problem can be formulated as a $(0, \infty)$-PC-ST: the terminals must be included (∞), while all other nodes are not relevant (0). In an analogous fashion, one can solve the $(0, \infty)$-PC-(path)-TSP on undirected graphs G_U, where the nodes with ∞ are the waypoints (and s, t).

As such, we can now apply all algorithmic and hardness results from the $(0, \infty)$ variant of the prize-collecting (path) TSP. However, the known results on the prize-collecting version of TSP are weaker than the ones of ΔTSP: this fact is not surprising, as PC-TSP is a generalization of its $(0, \infty)$ variant and ΔTSP.

The $(0, \infty)$-PC-TSP with $s = t$ may also be formulated as the Subset TSP, which asks for a (shortest) closed tour that visits a subset of nodes.

At this point one may wonder why to bother with the Subset TSP, given the parallels between UBWRP and the general metric TSP? As pointed out by Klein and Marx [29], the metric closure can destroy graph properties, e.g., planarity. For a more concrete example, consider the metric closure of a tree with unit link weights, removing unit weight properties. Hence, focusing on Subset TSP allows for algorithms with better approximation ratios on special graph classes.

Leveraging Subset TSP results for UBWRP. Klein and Marx [29] consider the Subset TSP problem as a cycle rather than a path. For planar graphs with k "waypoints", they give an algorithm with a runtime of $\left(2^{\mathcal{O}(\sqrt{k} \log k)} + W\right) \cdot n^{\mathcal{O}(1)}$, where W is the size of the largest link weight. Furthermore, they point out that the classic TSP dynamic programming techniques [6,20] can be applied to Subset TSP (with $s = t$), solving it optimally in a runtime of $2^k \cdot n^{\mathcal{O}(1)}$.

We show an extension to $s \neq t$, which enables us to use "cycle" algorithms as a black box: create a new node st, which serves as start and endpoint of the cycle, connecting st to two new waypoints w_s, w_t, and in turn w_s to s and w_t to t, where all new links have some arbitrarily large weight γ, see Fig. 4. W.l.o.g., we can assume that an optimal cycle solution of this modified graph starts with the path st, w_s, s. It is left to show that the subsequent tour ends with t, w_t, st: if not, the two nodes w_s, w_t would be traversed three times, which is a contradiction to optimality due to the choice of their incident link weight γ. Observe that instead of setting γ to "∞", $\gamma = W \cdot n^2$ suffices. Hence, we can use the algorithms forms [6,20,29] for the path version of subset TSP, and therefore, for UBWRP.

Corollary 7. UBWRP *can be solved optimally in a runtime of* $2^k \cdot n^{\mathcal{O}(1)}$ *for general graphs and in a runtime of* $\left(2^{\mathcal{O}(\sqrt{k} \log k)} + W\right) \cdot n^{\mathcal{O}(1)}$, *where* W *is the maximal link weight, for planar graphs.*

I.e., on general graphs, setting $k \in \mathcal{O}(\log n)$ is polynomial. For planar graphs, we can fix any $0 < x < 1$ with $k \in \mathcal{O}(\log^{1+x} n)$, $W \in n^{\mathcal{O}(1)}$ for polynomiality.

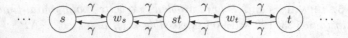

Fig. 4. Illustration of how to add the node st to the graph, connecting it to both s and t via w_s and w_t, respectively. Observe that a tour must traverse both w_s, w_t.

3   Ordered BWRP

The ordered version of BWRP turns out to be quite different in nature from the unordered one. First, we observe that while every BWRP instance is feasible, there are infeasible OBWRP instances due to capacity constraints. E.g., consider Fig. 5 with unit capacities and two waypoints. A special case is $k = 1$, which is identical to the unordered case, i.e., routing via one waypoint is always feasible and can be solved optimally in polynomial time.

Fig. 5. In this unit capacity network, the task is to route the flow of traffic from s to w_1, then to w_2, and lastly to t. To this end, the link from w_2 to w_1 must be used twice.

In the following, we study OBWRP in three contexts: (1) polynomial-time algorithms for computing feasible routes if $k \in \mathcal{O}(1)$ (Sect. 3.1), NP-hardness of optimality (Sect. 3.2), and optimality on cactus graphs, for any number of waypoints and constant capacities (Sect. 3.3). The latter is practically motivated by the fact that real computer networks often have specific topologies, and especially router-level wide-area topologies are usually quite sparse.

3.1   A Constant Number of Waypoints Is Feasible

There is a direct algorithmic connection from the link-disjoint path problem to OBWRP with unit capacities. By setting $s_1 = s$, $t_1 = w_1$, $s_2 = w_1$, $t_2 = w_2$, ..., a $k + 1$ link-disjoint path algorithm also solves unit capacity OBWRP for k waypoints. This method can be extended to general capacities via a standard technique, by replacing each link of capacity $c(e)$ with $\lfloor c(e) \rfloor$ parallel links of unit capacity and identical weight. To get rid of the parallel links, replace each link with a path of length two by "placing" a node on it, with the path weight being identical to the link weight.

Hence, we can apply the algorithm from Jarry and Prennes [22], which solves the feasibility of the link-disjoint path problem on bidirected graphs for a constant number of paths in polynomial runtime.

Theorem 8. *Let $k \in \mathcal{O}(1)$. Feasible solutions for OBWRP can be computed in polynomial time.*

The optimal solution already for few link-disjoint paths still puzzles researchers on bidirected graphs, but the problem seems to be non-trivial on undirected graphs as well: while feasibility for a constant number of link-disjoint paths is polynomial in the undirected case as well [25, 35], optimal algorithms for 3 or more link-disjoint paths are not known, and even for 2 paths the best result is a recent randomized high-order polynomial time algorithm [8]. For directed graphs, already 2 link-disjoint paths pose an NP-hard problem [17]. Results for waypoint routing on directed and undirected graphs are analogous [1].

3.2 Optimally Solving OBWRP Is NP-Hard

If we transition from a constant number to an arbitrary number of waypoints, we can show that then solving OBWRP optimally becomes NP-hard:

Theorem 9. *Solving OBWRP optimally is NP-hard.*

Proof. Reduction from the NP-hard link-disjoint path problem on bidirected graphs $G = (V, E)$ [10]: given k source-destination node-pairs (s_i, t_i), $1 \le i \le k$, are there k corresponding pairwise link-disjoint paths?

For every such instance I, we create an instance I' of OBWRP as follows, with all unit capacities: Set $s = s_1$ and $t = t_k$, also setting waypoints as follows: $w_1 = t_1$, $w_3 = s_2$, $w_4 = t_2$, $w_6 = s_3$, $w_7 = t_3$, ..., $w_{3k-3} = s_k$. We also create the missing $k-1$ waypoints $w_2, w_5, w_8, \ldots, w_{3k-4}$ as new nodes and connect them as follows, each time with bidirected links of weight γ: w_2 to $w_1 = t_1$ and $w_3 = s_2$, w_5 to $w_4 = t_2$ and $w_6 = s_3$, ..., w_{3k-4} to $w_{3k-3} = s_k$ and $w_{3k-5} = t_{k-1}$. I.e., we sequentially connect the end- and start-points of the paths.

Observe that OBWRP is feasible on I' if I is feasible: We take the k link-disjoint paths from I and connect them via the $k - 1$ new nodes in I'.

We now set γ to some arbitrarily high weight, e.g., $3k$ times the sum of all link weights. I.e., it is cheaper to traverse every link of I even $3k$ times rather than paying γ once. As thus, if I is feasible, the optimal solution of I' has a cost of less than $2 \cdot k \cdot \gamma$.

Assume I is not feasible, but that I' has a feasible solution $\mathscr{R}$. Observe that a feasible solution of I' needs to traverse the $k - 1$ new waypoints, i.e., has at least a cost of $2(k-1)\gamma$. As I was not feasible, we will now show that traversing every new waypoint $w_2, w_5, \ldots$ only once is not sufficient for a feasible solution of I'. Assume for contradiction that one traversal of $w_2, w_5, \ldots$ suffices: for each of those traversals of such a w_j, it holds that it must take place after traversing all waypoints with index smaller than j. Hence, we can show by induction that the removal of the links incident to the waypoints $w_2, w_5, \ldots$ from $\mathscr{R}$ contains a feasible solution for I. As thus, at least one of the waypoints $w_2, w_5, \ldots$ must be traversed twice, i.e., $\mathscr{R}$ has a cost of at least $2 \cdot k \cdot \gamma$.

We can now complete the polynomial reduction, by studying the cost (feasibility) of an optimal solution of I': if the cost is less than $2 \cdot k \cdot \gamma$, I is feasible, but if the cost is at least $2 \cdot k \cdot \gamma$ (or infeasible), I is not feasible.   □

While many OBWRP instances are not feasible (already in Fig. 5), we conjecture that the feasibility of OBWRP with arbitrarily many waypoints is NP-hard as well. This conjecture is supported by the fact that the analogous link-disjoint feasibility problems are NP-hard on undirected [18], directed [17], and bidirected graphs [10], also for undirected and directed ordered waypoint routing [1]. We thus turn our attention to special graph classes.

3.3   Optimality on the Cactus with Constant Capacity

The difficulty of OBWRP lies in the fact that the routing from w_i to w_{i+1} can be done along multiple paths, each of which could congest other waypoint connections. Hence, it is easy to solve OBWRP optimally (or check for infeasibility) on trees, as each path connecting two successive waypoints is unique.

Lemma 10. OBWRP *can be solved optimally in polynomial time on trees.*

For multiple path options, the problem turns NP-hard though (Theorem 9). To understand the impact of already two options, we follow-up by studying rings.

Lemma 11. OBWRP *is optimally solvable in polynomial time on bidirected ring graphs where for at least one link e holds:* $c(e) \in \mathcal{O}(1)$.

Proof. We begin our proof with $c(e) = c(e') = 1$. Observe that every routing between two successive waypoints has two path options P, clockwise or counterclockwise. We assign one arbitrary path P_e to traverse e, and another arbitrary path $P_{e'}$ to traverse e'. By removing the fully utilized e and e', the remaining graph is a tree with two leaves, where all routing is fixed, cf. Lemma 10.

We now count the path assignment possibilities for e, e': by also counting the "empty assignment", we have at most $(n+1)n$ options, where the optimal routing immediately follows for each option. For these $\mathcal{O}(n^2)$ possibilities, we pick the shortest feasible one. I.e., OBWRP can be solved optimally in polynomial time on rings with unit capacity. To extend the proof to constant capacities $c(e) \in \mathcal{O}(1)$, we use an analogous argument, the number of options for assignments to e and e' are now $\mathcal{O}\left(n^{2c(e)}\right) \in \mathrm{P}$. As thus, the lemma statement holds. $\square$

We now focus on the important case of cactus networks. Our empirical study using the Internet Topology Zoo[1] data set shows that over 30% are *cactus graphs*.

Theorem 12. OBWRP *is optimally solvable in polynomial time on cactus graphs with constant capacity.*

Proof. The idea is to (1) shrink the cactus graph down to a tree, (2) see if for the relevant subset of waypoints (to be described shortly) the feasibility holds on that tree, (3) reincorporate the excluded rings and find the optimal choice of path segments within each ring, and (4) construct an optimal route by stitching together the sub-routes obtained from the tree and the segments from each ring.

---

[1]   See http://www.topology-zoo.org/.

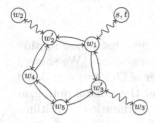 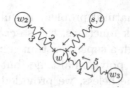 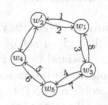

Fig. 6. In this cactus graph, we illustrate the algorithm of Theorem 12 w.r.t. the permutation $w_1 w_2 w_3 w_4 w_5$.

Fig. 7. Once the ring links are contracted, w' replaces the whole ring. Consequently, the permutation reduces to $w' w_2 w_3$. The sub-routes are numbered sequentially.

Fig. 8. The permutation induced on the ring is $w_1 w_2' w_3' w_4 w_5$. In the sub-problem, we have $s = t = w_1$. The numbers represent the order of node traversal in the optimal route.

Let $\mathcal{C}$ be the cactus graph (Fig. 6) and $T_{\mathcal{C}}$ be the tree obtained after contracting all the links on each rings. As a result of this link contraction, those waypoints previously residing on rings are now replaced by new (super) waypoints in $T_{\mathcal{C}}$ (Fig. 7). Each super node represents either a subtree of adjacent rings or just an isolated ring. Let $\mathcal{W}'$ denote the waypoints in $T_{\mathcal{C}}$.

Observe that any feasible route in $\mathcal{C}$ through $\mathcal{W}$ corresponds to one unique feasible route in $T_{\mathcal{C}}$ through nodes in $\mathcal{W}'$. Next, we show that either the feasible route in $T_{\mathcal{C}}$ (if exists) can be expanded to an optimal route for $\mathcal{C}$, or there is no feasible route in $\mathcal{C}$ at all. If $T_{\mathcal{C}}$ is not feasible then we are done. Otherwise, let $\mathcal{R}$ be the (unique) route in this tree. For each ring, $\mathcal{R}$ induces some *endpoints* (Fig. 8), one endpoint on each node that is either a) the joint of $T_{\mathcal{C}}$ and the ring, or b) the joint with its adjacent rings. Now we focus on the subproblem induced by this ring and the new waypoint set $\mathcal{W}''$ (to be specified) as follows.

For each endpoints that is visited by $\mathcal{R}$ add a waypoint to $\mathcal{W}''$. Then, using the algorithm described in the proof of Lemma 11, find an optimal route $\mathcal{R}_{ring}$ visiting all the nodes in $\mathcal{W}''$ respecting the order imposed by $\mathcal{R}$. If no such route exists, the instance is not feasible. Otherwise, remove from $\mathcal{R}$ every occurrence of the super node that represents this ring to get a disconnected route. For each missing part, reconnect the endpoints using the segment of $\mathcal{R}_{ring}$ restricted to these endpoints. Repeat this for every ring; denote the resulting route as $\mathcal{R}'$.

Finally, we argue that $\mathcal{R}'$ is optimal. This is the case because its pieces were taken from sets of sub-routes, where each set, covers a disjoint–or more precisely, vertex-disjoint up to endpoints–component of $\mathcal{C}$. Moreover, the set of sub-routes taken from an individual (disjoint) component (i.e. tree or ring) is optimal on that component. Therefore the total length is optimal. $\square$

4   Conclusion

We initiated the study of a natural problem in full-duplex networks: routing through a given set of network functions, the so-called waypoints. We showed that an optimal routing through a super-constant number of $\mathcal{O}(\log n)$ unordered waypoints can be computed in polynomial time, but that the general optimization problem is NP-hard. Nonetheless, we provided approximation algorithms with small constant competitive ratios for any number of waypoints, via connections to the Steiner Tree and (prize-collecting) Traveling Salesman problems. We also presented hardness results and polynomial-time algorithms for the ordered variant. In particular, we derived an exact polynomial-time algorithm for cactus graphs.

We believe that our work opens several interesting directions for future research. In general, while practically relevant, bidirected networks are not well-understood today, and assume an interesting position between directed and undirected networks. In particular, it would be interesting to understand for which bidirected graph classes the ordered and the unordered waypoint routing problem permits polynomial-time algorithms, and for up to how many waypoints. Another interesting direction for future research concerns the study of randomized algorithms.

Acknowledgements. Klaus-Tycho Foerster is supported by VILLUM FONDEN project ReNet and Mahmoud Parham by AAU's PreLytics project.

References

1. Amiri, S.A., Foerster, K.-T., Jacob, R., Schmid, S.: Charting the Complexity Landscape of Waypoint Routing. arXiv preprint arXiv:1705.00055 (2017)
2. Amiri, S.A., Foerster, K.-T., Schmid, S.: Walking through waypoints. arXiv preprint arXiv:1708.09827 (2017)
3. Archer, A., Bateni, M.H., Hajiaghayi, M.T., Karloff, H.J.: Improved approximation algorithms for prize-collecting steiner tree and TSP. SIAM J. Comput. **40**(2), 309–332 (2011)
4. Arora, S., Grigni, M., Karger, D.R., Klein, P.N., Woloszyn, A.: A polynomial-time approximation scheme for weighted planar graph TSP. In: Proceedings of SODA (1998)
5. Bansal, N., Lee, K.-W., Nagarajan, V., Zafer, M.: Minimum congestion mapping in a cloud. In: Proceedings of PODC (2011)
6. Bellman, R.: Dynamic programming treatment of the travelling salesman problem. J. ACM **9**(1), 61–63 (1962)
7. Björklund, A., Husfeld, T., Taslaman, N.: Shortest cycle through specified elements. In: Proceedings of SODA (2012)
8. Björklund, A., Husfeldt, T.: Shortest two disjoint paths in polynomial time. In: Esparza, J., Fraigniaud, P., Husfeldt, T., Koutsoupias, E. (eds.) ICALP 2014. LNCS, vol. 8572, pp. 211–222. Springer, Heidelberg (2014). https://doi.org/10.1007/978-3-662-43948-7_18
9. Byrka, J., Grandoni, F., Rothvoß, T., Sanità, L.: Steiner tree approximation via iterative randomized rounding. J. ACM **60**(1), 6:1–6:33 (2013)

10. Chanas, P.: Reseaux ATM: conception et optimisation. Ph.D. thesis, University of Grenoble (1998)
11. Christofides, N.: Worst-case analysis of a new heuristic for the travelling salesman problem. Technical report 388, Graduate School of Industrial Administration, Carnegie Mellon University (1976)
12. Cormen, T.H., Leiserson, C.E., Rivest, R.L., Stein, C.: Introduction to Algorithms, 3rd edn. MIT Press, Cambridge (2009)
13. Edmonds, J., Johnson, E.: Matching: a well-solved class of linear programs. In: Combinatorial Structures and their Applications: Proceedings of the Calgary Symposium, pp. 88–92. Gordon and Breach, New York (1970)
14. ETSI: Network functions virtualisation - introductory white paper. White Paper, October 2013
15. Even, G., Medina, M., Patt-Shamir, B.: On-line path computation and function placement in SDNs. In: Bonakdarpour, B., Petit, F. (eds.) SSS 2016. LNCS, vol. 10083, pp. 131–147. Springer, Cham (2016). https://doi.org/10.1007/978-3-319-49259-9_11
16. Even, G., Rost, M., Schmid, S.: An approximation algorithm for path computation and function placement in SDNs. In: Suomela, J. (ed.) SIROCCO 2016. LNCS, vol. 9988, pp. 374–390. Springer, Cham (2016). https://doi.org/10.1007/978-3-319-48314-6_24
17. Fortune, S., Hopcroft, J.E., Wyllie, J.: The directed subgraph homeomorphism problem. Theor. Comput. Sci. **10**, 111–121 (1980)
18. Garey, M.R., Johnson, D.S.: Computers and Intractability: A Guide to the Theory of NP-Completeness. W. H. Freeman, New York (1979)
19. Hartert, R., Vissicchio, S., Schaus, P., Bonaventure, O., Filsfils, C., Telkamp, T., Francois, P.: A declarative and expressive approach to control forwarding paths in carrier-grade networks. In: Proceedings of SIGCOMM (2015)
20. Held, M., Karp, R.M.: The traveling-salesman problem and minimumspanning trees: Part II. Math. Program. **1**(1), 6–25 (1971)
21. Sherry, J., et al.: Making middleboxes someone else's problem: network processing as a cloud service. In: Proceedings of ACM SIGCOMM (2012)
22. Jarry, A., Pérennes, S.: Disjoint paths in symmetric digraphs. Discrete Appl. Math. **157**(1), 90–97 (2009)
23. Jarry, A.: Multiflows in symmetric digraphs. Discrete Appl. Math. **160**(15), 2208–2220 (2012)
24. Karpinski, M., Lampis, M., Schmied, R.: New inapproximability bounds for TSP. J. Comput. Syst. Sci. **81**(8), 1665–1677 (2015)
25. Kawarabayashi, K., Kobayashi, Y., Reed, B.A.: The disjoint paths problem in quadratic time. J. Comb. Theor. Ser. B **102**(2), 424–435 (2012)
26. Khuller, S., Mitchell, S.G., Vazirani, V.V.: Processor efficient parallel algorithms for the two disjoint paths problem and for finding a kuratowski homeomorph. SIAM J. Comput. **21**(3), 486–506 (1992)
27. Khuller, S., Schieber, B.: Efficient parallel algorithms for testing k-connectivity and finding disjoint s-t paths in graphs. SIAM J. Comput. **20**(2), 352–375 (1991)
28. Klein, P.N.: A subset spanner for planar graphs, with application to subset TSP. In: Proceedings of STOC (2006)
29. Klein, P.N., Marx, D.: A subexponential parameterized algorithm for subset TSP on planar graphs. In: Proceedings of SODA (2014)

30. Lukovszki, T., Schmid, S.: Online admission control and embedding of service chains. In: Scheideler, C. (ed.) Structural Information and Communication Complexity. LNCS, vol. 9439, pp. 104–118. Springer, Cham (2015). https://doi.org/10.1007/978-3-319-25258-2_8

31. Napper, J., Haeffner, W., Stiemerling, M., Lopez, D.R., Uttaro, J.: Service Function Chaining Use Cases in Mobile Networks. Internet-draft, IETF (2016)

32. Naves, G., Sebö, A.: Multiflow feasibility: an annotated tableau. In: Cook, W., Lovász, L., Vygen. J. (eds.) Research Trends in Combinatorial Optimization, pp. 261–283. Springer, Heidelberg (2009). https://doi.org/10.1007/978-3-540-76796-1_12

33. Sköldström, P., et al.: Towards unified programmability of cloud and carrier infrastructure. In: Proceedings of EWSDN (2014)

34. Soulé, R., et al.: Merlin: a language for provisioning network resources. In: Proceedings of ACM CoNEXT (2014)

35. Robertson, N., Seymour, P.D.: Graph minors .XIII. The disjoint paths problem. J. Comb. Theor. Ser. B 63(1), 65–110 (1995)

36. Robins, G., Zelikovsky, A.: Tighter bounds for graph steiner tree approximation. SIAM J. Discrete Math. 19(1), 122–134 (2005)

37. Saltzer, J.H., Reed, D.P., Clark, D.D.: End-to-end arguments in system design. ACM Trans. Comput. Syst. (TOCS) 2(4), 277–288 (1984)

38. Sebö, A., van Zuylen, A.: The salesman's improved paths: A 3/2+1/34 approximation. In: Proceedings of FOCS (2016)

39. Vachani, R., Shulman, A., Kubat, P., Ward, J.: Multicommodity flows in ring networks. INFORMS J. Comput. 8(3), 235–242 (1996)

Service Chain Placement in SDNs

Gilad Kutiel[1,2]($\boxtimes$) and Dror Rawitz[1,2]

[1] Department of Computer Science, Technion, 32000 Haifa, Israel
gkutiel@cs.technion.ac.il
[2] Faculty of Engineering, Bar Ilan University, 52900 Ramat-Gan, Israel
dror.rawitz@biu.ac.il

Abstract. We study the allocation problem of a compound request for a *service chain* in a software defined network that supports network function virtualization. Given a network that contains servers with limited processing power and of links with limited bandwidth, a service chain is a sequence of virtual network functions (VNFs) that service a certain flow request in the network. The allocation of a service chain consists of routing and VNF placement, namely each VNF from the sequence is placed in a server along a path. It is feasible if each server can handle the VNFs that are assigned to it, and if each link on the path can carry the flow that is assigned to it. A request for service is composed of a source and a destination in the network, an upper bound on the total latency, and a specification in the form of a directed acyclic graph (DAG) of VNFs that provides all service chains that are considered valid for this request. In addition, each pair of server and VNF is associated with a cost for placing the VNF in the server. Given a request, the goal is to find a valid service chain of minimum total cost that respects the latency constraint or to identify that such a service chain does not exist.

We show that even the feasibility problem is NP-hard in general graphs. Hence we focus on DAGs. We show that the problem is still NP-hard in DAGs even for a very simple network, and even if the VNF specification consists of only one option (i.e., the virtual DAG is a path). On the other hand, we present an FPTAS for the case where the network is a DAG. In addition, based on our FPTAS, we provide algorithms for instances in which the service chain passes through a bounded number of vertices whose degree is larger than two.

1 Introduction

Computer communication networks are in constant need of expansion to cope with the ever growing traffic. As networks grow, management and maintenance become more and more complicated. Current developments that aim to improve the utilization of network resources include the detachment of network applications from network infrastructure and the transition from network planning to network programming.

Research supported in part by Network Programming (Neptune) Consortium, Israel.
D. Rawitz—Partially supported by the Israel Science Foundation (grant no. 497/14).

© Springer International Publishing AG, part of Springer Nature 2018
D. Alistarh et al. (Eds.): ALGOCLOUD 2017, LNCS 10739, pp. 27–40, 2018.
https://doi.org/10.1007/978-3-319-74875-7_3

One aspect of network programming is to manage resources from a central point of view, namely to make decisions based on availability, network status, required quality of service, and the identity of the client. Hence, a focal issue is a central agent that is able to received reports from network components and client requests, and as a result can alter the allocation of resources in the networks. This approach is called Software Defined Networking (SDN), where there is a separation between routing and management (control plane) and the underlying routers and switches that forward traffic (data plane) (see Kreutz et al. [11]).

A complementary approach is Network Function Virtualization (NFV) [3]. Instead of relying on special purpose machines, network applications become virtual network functions (VNF) that are executed on generic machines and can be placed in various locations in the network. Virtualization increases the flexibility of resource allocation and thus the utilization of the network resources. Internet Service Providers (ISPs) that provide services to clients benefit from NFV, since it helps to better utilize their physical network. In addition, when network services are virtualized, an ISP may support *service chaining* [1], namely a compound service that consists of a sequence of VNFs. Furthermore, one may offer a compound service a service that may be obtained using one of several sequences of VNFs.

1.1 Related Work

In this paper we consider an SDN network that employs NVF [4]. Such networks attracted a lot of attention in the networking community, especially from a systems point of view (see, e.g., [9,10]). Several papers also consider the algorithmic aspects of such networks but mainly present heuristics (see, e.g., [15]). For more details see [6] and references therein.

Cohen et al. [2] considered VNF placement. In their model the input is an undirected graph with a metric distance function on the edges. Clients are located in various nodes, and each client is interested in a subset of VNFs. For each node v and VNF α, there is a setup cost associated with placing a copy of α at v (multiple copies are allowed), and there is a load that is induced on v for placing a copy of α. Furthermore, each node has a limited capacity, and each copy of a VNF can handle a limited amount of clients. A solution is the assignment of each client to a subset of nodes, each corresponding to one of its required VNFs. The cost of a solution is the total setup costs plus the sum of distances between the clients and the node from which they get service. Cohen et al. [2] gave bi-criteria approximation algorithms for various versions of the problem, namely algorithms that compute constant factor approximations that violate the load constraints by a constant factor. It is important to note that in this problem routing is not considered and it is assumed that VNF subsets can be executed in any order.

Lukovszki and Schmid [12] studied the problem of admission control and placement of service chains, where each chain is associated with a source-destination pair and is supposed to be routed via an ordered sequence of ℓ VNFs. The VNFs may have multiple instantiations, but each node has a limited capacity that bounds the number of requests it may service. They presented an $O(\log \ell)$-competitive algorithm for the problem of maximizing the number of serviced chains, assuming that capacities are $\Omega(\log \ell)$. It is also shown that this ratio is asymptotically optimal even for randomized online algorithms. APX-hardness results for the offline version of the problem were also presented. Rost and Schmid [13] considered variant of this problem, where each node can host a subset of the VNFs. In addition, each VNF has a demand and each VNF-node pair has a capacity, and in a feasible solution the total demand for each pair is bounded by the capacity. They considered two goals: maximum profit and minimum resource cost and gave bicriteria approximation algorithms which are based on LP-rounding for several special cases.

Even et al. [5] studied online path computation and VNF placement. They considered compound requests that arrive in an online manner. Each request is a flow with a specification of routing and VNF requirements which is represented by a directed acyclic graph (DAG) whose vertices are VNFs. Each VNF can be performed by a specified subset of servers in the system. Upon arrival of a request, the algorithm either rejects the request or accepts it with a specific routing and VNF assignment, under given server capacity constraints. Each request has a benefit that is gained if it is accepted, and the goal is to maximize the total benefit gained by accepted requests. Even et al. [5] proposed an algorithm that copes with requests with unknown duration without preemption by using a third response, which is refer to as "stand by", whose competitive ratio is $O(\log(knb_{\max}))$, where n is the number of nodes, $b_{\max}$ is an upper bound on a request benefit per time unit, and k is the maximum number of VNFs of any service chain. This performance guarantee holds for the case where the processing of any request in any possible service chain is never more than $O(1/(k \log(nk)))$ fraction of the capacity of any network component. Even et al. [6] presented a randomized approximation algorithm for the same problem that computes $(1 - \varepsilon)$-approximate placements if the demand of any processing request in any possible service chain is never more than an $O(\varepsilon^2/\log n)$ fraction of the capacity of any network component.

1.2   Our Results

We study the allocation problem of a compound flow request for a service chain in a software-defined network that supports NFV, where the goal is to find an optimal placement with respect to the current available resources from the point of view of the network (i.e., the ISP). More specifically, we consider the case where the network already contains previous resource allocations. Therefore, we are given a (physical) network that contains servers with limited (residual) processing power and of links with limited (residual) bandwidth. A request for service is composed of a source and a destination in the network, an upper bound on the total latency, and a specification of all service chains that are considered valid for this request. As in [5], the specification is represented by a DAG whose vertices are VNFs. The allocation of a service chain consists of routing and VNF placement. That is, each VNF from the sequence is placed in a server along a path, and it is feasible if each server can handle the VNFs that are assigned to it, and if each link on the path can carry the flow that is assigned to it. Moreover each link causes a delay, and the service chain placement should also comply with a global bound on the total latency. Each pair of server and VNF is associated with a cost for placing the network function in the server. This cost measures compatibility of a VNF to a server (e.g., infinite costs means "incompatible"). Given a request, the goal is to find a feasible service chain of minimum total cost or to identify that a valid service chain does not exist.

We show that even feasibility is NP-hard in general networks using a reduction from HAMILTONIAN PATH. We show that the problem is still NP-hard if the network is a DAG even for a very simple network, and even if the specification consists of a single option. Both reductions are from PARTITION.

On the positive side, we present an FPTAS for the case where the network is a DAG which is based on a non-trivial dynamic programming algorithm. Based on our FPTAS, we provide a randomized algorithm for general networks in which there is an optimal placement whose physical path consists of at most k vertices whose degree is larger than 2. For example, this can be assumed if all simple paths from s to t contain at most k such vertices. The algorithm computes a $(1 + \varepsilon)$-approximate placement with high probability using $k! \log n$ invocations of the FPTAS. We also present a (deterministic) parameterized algorithm that computes a $(1 + \varepsilon)$-approximate placement in time $O(k! \cdot \mathrm{poly}(n))$, where k is the number of vertices in the network whose degree is larger than 2.

2   Preliminaries

Model. An instance of the SERVICE CHAIN PLACEMENT (SCP) problem is composed of three components, a physical network, a virtual specification, and placement costs:

Physical network: The physical network is a graph $G = (V, E)$. Each node $v \in V$ has a non-negative processing capacity $p(v)$, and each directed edge $e \in E$ has a non-negative bandwidth capacity $b(e)$. Without loss of generality, we assume that $p(s) = p(t) = 0$.

Virtual specification: The description of a request for a service chain consists of a physical source $s \in V$, a physical destination $t \in V$, and a directed acyclic graph (DAG) $\mathcal{G} = (\mathcal{V}, \mathcal{E})$. The DAG $\mathcal{G}$ has a source node $\sigma \in \mathcal{V}$ and a destination $\tau \in \mathcal{V}$. Each node $\alpha \in \mathcal{V}$ represent a VNF that has a processing demand $p(\alpha)$. Without loss of generality, we assume that $p(\sigma) = p(\tau) = 0$. Each edge $\varepsilon \in \mathcal{E}$ has a bandwidth demand $b(\varepsilon)$.

Placement costs: There is a non-negative cost $c(\alpha, v)$ for placing the VNF α in v. We assume that $c(\sigma, s) = 0$ and $c(\sigma, v) = \infty$, for every $v \neq s$. Similarly, we assume that $c(\tau, t) = 0$ and $c(\tau, v) = \infty$, for every $v \neq t$.

A solution consists of the following:

Virtual path: A path from σ to τ in $\mathcal{G}$, namely a sequence of vertices $\alpha_0, \ldots, \alpha_q$, where $\alpha_0 = \sigma$, $\alpha_q = \tau$, and $(\alpha_j, \alpha_{j+1}) \in \mathcal{E}$, for every $j \in \{0, \ldots, q-1\}$.

Physical path: A simple path from s to t in G, namely a sequence of nodes $v_0, \ldots, v_k$, where $v_0 = s$, $v_k = t$, and $(v_i, v_{i+1}) \in E$, for every $i \in \{0, \ldots, k-1\}$.

Placement: A function f that maps a virtual path to a simple physical path. Formally, a placement is a function $f : \{\alpha_0, \ldots, \alpha_q\} \rightarrow \{v_0, \ldots, v_k\}$ where

1. $v_i \neq v_{i'}$ if $i \neq i'$.
2. If $f(\alpha_j) = v_i$ and $f(\alpha_{j+1}) = v_{i'}$, then $i \leq i'$.
3. $b(\alpha_j, \alpha_{j+1}) \leq b(v_i, v_{i+1})$, for every $(v_i, v_{i+1}) \in \bar{f}(\alpha_j, \alpha_{j+1})$, where $\bar{f}(\alpha_j, \alpha_{j+1})$ to be the set of physical edges that correspond to it, i.e.,

$$\bar{f}(\alpha_j, \alpha_{j+1}) \triangleq \{(v_i, v_{i+1}) : i' \leq i < i''\},$$

where $f(\alpha_j) = v_{i'}$ and $f(\alpha_{j+1}) = v_{i''}$.
4. $\sum_{\alpha \in f^{-1}(v)} p(\alpha) \leq p(v)$, where $f^{-1}(v) \triangleq \{\alpha : f(\alpha) = v\}$.

An example of an SCP instance and a solution is given in Fig. 1.

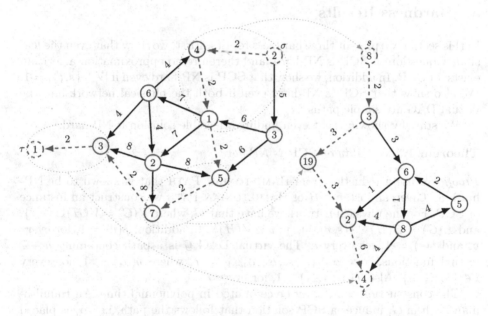

Fig. 1. An example of an SCP instance and a solution.

The cost of a placement f is defined as:

$$c(f) \triangleq \sum_j c(\alpha_j, f(\alpha_j)) .$$

In SCP the goal is to find a feasible placement of a virtual path into the physical DAG that minimizes the cost.

We also consider an extended version of SCP in which each physical link causes a delay and there is an upper bound L on the total delay. More formally, each pair of a virtual edge ε and physical edge e is associated with a latency $\ell(\varepsilon, e)$. Given a placement $f : \{\alpha_0, \ldots, \alpha_q\} \rightarrow \{v_0, \ldots, v_k\}$, the total latency of the solution is given by

$$L(f) = \sum_{j=0}^{q-1} \sum_{e \in \bar{f}((\alpha_j, \alpha_{j+1}))} \ell((\alpha_j, \alpha_{j+1}), e) .$$

In this case there is an additional constraint that $L(f) \leq L$.

Notation. Given a virtual DAG $\mathcal{G}$, we assume the existence of a topological sorting of the vertices and write $\alpha \prec \beta$ if α precedes β in this ordering. If the physical network is a DAG, then we make a similar assumption and write $v \prec u$ if v precedes u in this ordering.

3 Hardness Results

In this section we present three hardness results. First, we show that even the feasibility question of SCP is NP-hard, and therefore no approximation algorithm exists for SCP. In addition, we show that SCP is NP-hard even if $|V \setminus \{s, t\}| = 1$. We also show that SCP is NP-hard even if both the physical network and the virtual DAG are simple paths.

We start by showing that even finding a feasible solution is NP-hard.

Theorem 1. *Feasibility of* SCP *is NP-hard*

Proof. We use a reduction from HAMILTONIAN PATH that is known to be NP-hard [8]. Given an instance H of HAMILTONIAN PATH we construct an instance of SCP. For the physical network we have that G, where $V(G) = V(H) \cup \{s, t\}$, and $E(G) = E(H) \cup \{(s, v), (v, t) : v \in V(H)\}$. In addition, $p(v) = 1$, for every v, and $b(e) = 1$, for every e. The virtual DAG $\mathcal{G}$ is a path containing $n + 2$ virtual functions, $\sigma = \alpha_0, \alpha_1, \ldots, \alpha_n, \alpha_{n+1} = \tau$, where $p(\alpha_i) = 1$, for every $i \in \{1, \ldots, n\}$. Also, $b(\alpha_i, \alpha_{i+1}) = 1$, for every i.

The construction can clearly be computed in polynomial time. An Hamiltonian Path in G, induces a SCP solution that follows the path, i.e., α_i is placed in the ith vertex along the path. On the other hand, since all demands and capacities are 1, no two VNFs can share a physical node. Hence a SCP solution induces an Hamiltonian path.   □

Next, we show that SCP is NP-hard even if the physical network is very simple.

Theorem 2. SCP *is NP-hard, even if* $|V \setminus \{s, t\}| = 1$.

Proof. We prove the theorem using a reduction from PARTITION. Given a PARTITION instance $\{a_1, \ldots, a_n\}$, we construct an SCP instance as follows. The physical network G contains three nodes: s, v, and t, and there are two edges (s, v) and (v, t). The capacity of v is $p(v) = \frac{1}{2}\sum_i a_i$. Edge bandwidth are zero. The virtual DAG is composed of $2n + 2$ vertices, namely

$$\mathcal{V} = \{\sigma, \tau\} \cup \{\alpha_1, \ldots, \alpha_n\} \cup \{\beta_1, \ldots, \beta_n\}.$$

Also,

$$\mathcal{E} = \bigcup_i \{\{\alpha_i, \beta_i\} \times \{\alpha_{i+1}, \beta_{i+1}\}\} \cup \{(\sigma, \alpha_1), (\sigma, \beta_1), (\alpha_n, t), (\beta_n, t)\}.$$

The DAG is shown in Fig. 2. The demands are $p(\alpha_i) = a_i$ and $p(\beta_i) = 0$, for every i. Also, $b(\varepsilon) = 0$, for every $\varepsilon \in \mathcal{E}$. The costs are: $c(\alpha_i, v) = 0$ and $c(\beta_i, v) = a_i$, for every i.

The SCP instance can be computed in polynomial time. Also, it is not hard to verify that $\{a_1, \ldots, a_n\} \in$ PARTITION if and only if there exists a solution whose cost is $\frac{1}{2}\sum_i a_i$. □

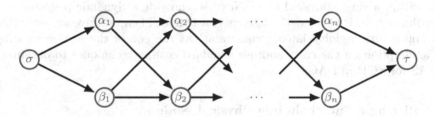

Fig. 2. The specification defined in the proof of Theorem 2.

Next, we show that SCP is NP-hard even if both the physical network and the VNF specification are paths.

Theorem 3. SCP *is NP-hard, even if both the physical network and the virtual DAG are paths.*

Proof. We prove the theorem using a reduction from PARTITION. Given a PARTITION instance $\{a_1, \ldots, a_n\}$, we construct a virtual path $\sigma, \alpha_1, \ldots, \alpha_n, \tau$ and a physical path $s, v_1^-, v_1^+, \ldots, v_n^-, v_n^+, t$. We set $p(v) = 1$, for every node $v \neq s, t$, and $p(\alpha_i) = 1$, for every i. In addition, we set $b(e) = 1$, for every edge $e \in E$, and $b(\varepsilon) = 1$, for every edge $\varepsilon \in \mathcal{E}$. As for the costs, we define $c(\alpha_i, v_i^-) = 0$, $c(\alpha_i, v_i^+) = a_i$, and for any $v \notin \{v_i^-, v_i^+\}$ we set $c(\alpha_i, v) = \infty$. We also define $\ell((\alpha_i, \alpha_{i+1}), (v_i^-, v_i^+)) = a_i$, and set $\ell(\varepsilon, e) = 0$, otherwise. Finally, we set $L = \frac{1}{2}\sum_i^n a_i$. Figure 3 depicts the above reduction. One can verify that a_i is either counted in the latency of in the cost. Hence, $\{a_1, \ldots, a_n\} \in$ PARTITION if and only if there is a placement with cost $\frac{1}{2}\sum_i^n a_i$. □

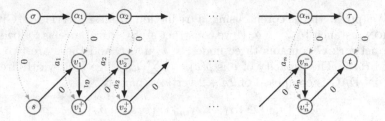

Fig. 3. A reduction from a partition instance $a_1, \ldots, a_n$. The cost function enforces each α_i to be placed either in v_i^- or in v_i^+. In the former case this placement incure no cost but additional latency of a_i, in the later case there will be additional a_i cost with zero latency.

4   Algorithms for Physical Directed Acyclic Graphs

In this section we present an FPTAS for SCP in DAGs which is based on a dynamic programming algorithm. The algorithm is described in a top down manner. We first assume that costs are polynomially bounded and design a dynamic programming algorithm that computes a minimum cost placement of a virtual path within a single physical node. Next, we provide a dynamic programming algorithm for SCP without the latency constraint. Then, we give an algorithm that copes with a global latency constraint. At the end of the section we drop our assumption on the costs and use standard scaling techniques to obtain an FPTAS for SCP in DAGs.

4.1   Placing a Sub-chain in a Physical Node

Assume that we want to place a minimum cost virtual path from a VNF α to a VNF β within a physical node v. A sequence of VNFs $\alpha = \alpha_0, \alpha_2, \ldots, \alpha_q = \beta$ is a *candidate path* if $(\alpha_i, \alpha_{i+1}) \in \mathcal{E}$, for every i, and $\sum_i p(\alpha_i) \leq p(v)$. We would like to find such a path with minimum cost, and we denote the cost of such a path by $\mathrm{COST}_v(\alpha \rightsquigarrow \beta)$.

We use dynamic programming in order to compute such a path. Let $\mathrm{PRC}_v(\alpha \rightsquigarrow \beta, c)$ be the minimum amount of processing required to place a virtual path form α to β into v among paths whose cost is at most c. The value of $\mathrm{PRC}_v(\alpha \rightsquigarrow \beta, c)$ can be computed recursively as follow:

$$
\mathrm{PRC}_v(\alpha \rightsquigarrow \beta, c) = \begin{cases} \infty & c(\alpha, v) > c, \\ p(\alpha) & \begin{array}{l} c(\alpha, v) \leq c, \\ \beta = \alpha, \end{array} \\ \min_{(\gamma, \beta) \in \mathcal{E}} \{\mathrm{PRC}_v(\alpha \rightsquigarrow \gamma, c - c(\beta, v)) + p(\beta)\} & \text{otherwise.} \end{cases}
$$

Observe that if $c(\alpha, v) > c$, then a placement is not possible with a budget c. Otherwise, if $\alpha = \beta$, then the best path is the one containing only α. If $\alpha \neq \beta$, then an optimal path ends with an edge $(\gamma, \beta) \in \mathcal{E}$, which means that the processing placed at v consists of $p(\beta)$ plus the minimum amount of processing of a path from α to γ, whose cost is at most $c - c(\beta, v)$.

To complete our argument observe that the following holds:

$$\text{COST}_v(\alpha \rightsquigarrow \beta) = \min\left\{c : \text{PRC}_v(\alpha \rightsquigarrow \beta, c) \leq p(v)\right\}.$$

Since the costs are assumed to be polynomially bounded, there is a polynomial number of states to be computed, and the computation of each of them can be done in linear time. Hence, the total running time is polynomial. Finally, we note that the above algorithm computes the minimum amount of processing, but may also be used to compute the actual placement that achieves this value using standard techniques.

4.2 Placing a Service Chain

In this section we describe a dynamic programming algorithm for placing a service chain without a global latency bound.

Consider an optimal solution of SCP, i.e., a service chain from σ to τ which is placed along a path from s to t in the physical DAG. Suppose v is a node along the path from s to t. Also, let α be the last VNF in the service chain which is placed in the path from s to v. It must be that the placement of the VNFs from σ to α in the path from s to v is the best one among all placements of a virtual path from σ to α in a path from s to v. In other words, any partial placement of an optimal placement is also optimal. Our algorithm is based on this property.

We define a state for each pair of VNF α and physical node v. Let $\text{COST}(\alpha, v)$ stand for the minimum cost placement of a virtual path from σ to α in a path from s to v, where α is the last VNF which is placed along the physical path. In addition, we write $(u, v) \mapsto (\gamma, \beta)$ if v is reachable from u using only edges with bandwidth at least $b(\gamma, \beta)$, namely if there is a path from u to v such that the bandwidth of all edges in the path is at least $b(\gamma, \beta)$. $\text{COST}(\alpha, v)$ can be computed recursively, as follows:

$$\text{COST}(\alpha, v) = \begin{cases} 0 & \alpha = \sigma, v = s, \\ \min\limits_{\substack{\beta \prec \alpha, (\gamma, \beta) \in \mathcal{E}, \\ u \prec v, \\ (u,v) \mapsto (\gamma, \beta)}} \left\{\text{COST}(\gamma, u) + \text{COST}_v(\beta \rightsquigarrow \alpha)\right\} & \text{otherwise.} \end{cases}$$

The desired value is $\text{COST}(\tau, t)$. Consider an optimal placement of a virtual path ending at α within a path ending at v. Let γ be the first VNF along the virtual path that is not placed in v. Also, let u be the node in which γ is placed. Hence, the optimal placement is composed of three segments: an optimal placement of a virtual path that ends in γ in a physical path ending at u, a virtual edge (γ, β) is placed in the path from u to v, and a minimum cost placement of a virtual path from β to α in v, that does not violate the capacity of v. Thus, we check all pairs (γ, u), where $\gamma \prec \alpha$ and $u \prec v$, and for each pair we consider any neighbor

Virtual Graph

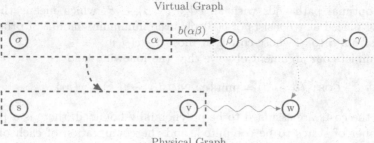

Physical Graph

Fig. 4. The optimal placement of a path to α into a path to v can be efficiently computed by breaking the problem into the problem of placing a path to γ-path into a path to u (blue, dashed rectangles) and the problem of embedding a path from β to α into v (orange, dotted). The path from u to v must carry at least $b(\gamma, \beta)$ bandwidth. (Color figure online)

$(\gamma, \beta) \in \mathcal{E}$ and $\beta \prec \alpha$ such that $(u, v) \Mapsto (\gamma, \beta)$. The recursive computation is illustrated in Fig. 4.

As for the running time, observe that checking whether $(u, v) \Mapsto (\gamma, \beta)$ can be done using DFS in linear time. Since the number of state is polynomial, and each state can be computed in polynomial time, the total running time is polynomial. Finally, we note that the above algorithm computes the minimum amount of cost, but may also be used to compute the actual placement that achieves this value using standard techniques.

4.3 Placing a Service Chain with a Latency Bound

We now consider the SCP problem with latency. Recall that in this variant of the problem we are also given a latency function $\ell : E \times \mathcal{E} \to \mathbb{R}_+$, and a latency upper bound L. The goal is to find a minimum cost placement that also respect the latency constraint.

Let $\mathrm{LAT}(\alpha, v, c)$ be the minimum latency created by a placement of a virtual path from σ to α in a path from s to v whose cost is at most c. Also, let $\mathrm{LAT}(\gamma, \beta, w \rightsquigarrow v)$ be the minimum possible latency of path from w to v such that the bandwidth of all edges in the path is at least $b(\gamma, \beta)$. That is,

$$\mathrm{LAT}(\gamma, \beta, w \rightsquigarrow v) = \min_{f : b(\gamma, \beta) \leq \min_{e \in \bar{f}(\gamma, \beta)} b(e)} \sum_{e \in \bar{f}(\gamma, \beta)} \ell((\gamma, \beta), e) \,.$$

Observe that $\mathrm{LAT}(\gamma, \beta, w \rightsquigarrow v)$ can be computed using a SHORTEST PATH algorithm, were we consider only edges whose bandwidth cap is at least $b(\gamma, \beta)$. Now we are ready for the computation of $\mathrm{LAT}(\alpha, v, c)$:

$$\text{LAT}(\alpha, v, c) = \begin{cases} \infty & c < 0, \\ 0 & \begin{aligned} &\alpha = \sigma, v = s, \\ &\text{and } c \geq 0, \end{aligned} \\ \min\limits_{\substack{\beta \prec \alpha, (\gamma, \beta) \in \mathcal{E}, \\ w \prec v, (u,w) \in E, \\ (u,v) \mapsto (\gamma, \beta)}} \left\{ \begin{aligned} &\text{LAT}(\gamma, \beta, w \rightsquigarrow v) + \\ &\text{LAT}(\gamma, u, c - \text{COST}_v(\beta \rightsquigarrow \alpha)) \end{aligned} \right\} & \text{otherwise.} \end{cases}$$

The minimum latency of a placement of a path to α in a path to v whose cost is at most c can be divided into two values: the latency caused by the placement of a path to γ, where $(\gamma, \beta) \in \mathcal{E}$, and $\beta \prec \alpha$, with cost $c - \text{COST}_v(\beta \rightsquigarrow \alpha)$, and the latency caused by placing a virtual edge (γ, β) on a physical path from w to v, where $w \prec v$ and $(u, w) \in E$. The optimal value is $\min\{c : \text{LAT}(\tau, t, c) \leq L\}$.

Since the computation of $\text{LAT}(\gamma, \beta, w \rightsquigarrow v)$ can be done efficiently, the computation for a triple (α, v, c) can be done in polynomial time. Hence, the total running time is polynomial. Finally, as in the previous algorithms, one may use the algorithm to compute a corresponding placement using standard techniques.

4.4 FPTAS for General Costs

In this section we present an FPTAS for SCP in DAGs for general costs, namely without the assumption that costs are polynomially bounded. Our algorithm is similar to the FPTAS for the MINIMUM KNAPSACK problem.

Let f be an optimal placement, and let $c_{\max}$ be the maximum cost of a VNF placement by f, i.e., $c_{\max} = \max_{f(\alpha)=u} c(\alpha, v)$. Given a constant $\varepsilon > 0$, we define

$$c'(\alpha, v) \triangleq \begin{cases} \left\lceil \frac{c(\alpha,v)}{c_{\max}} \cdot \frac{n}{\varepsilon} \right\rceil & c(\alpha, v) \leq c_{\max}, \\ \infty & c(\alpha, v) > c_{\max}. \end{cases}$$

Let f' be an optimal placement with respect to c'.

Lemma 1. $c(f') \leq (1 + \varepsilon)c(f)$.

Proof. We have that

$$c(f') = \sum_\alpha c(\alpha, f'(\alpha)) \leq \frac{\varepsilon c_{\max}}{n} \cdot \sum_\alpha c'(\alpha, f'(\alpha))$$

$$\leq \frac{\varepsilon c_{\max}}{n} \cdot \sum_\alpha c'(\alpha, f(\alpha))$$

$$\leq \frac{\varepsilon c_{\max}}{n} \cdot \sum_\alpha \left(c(\alpha, f(\alpha)) \cdot \frac{\varepsilon c_{\max}}{n} + 1 \right)$$

$$= c(f) + \varepsilon c_{\max}$$

$$\leq (1 + \varepsilon)c(f),$$

where the second inequality is due to the optimality of f' with respect to c'. $\square$

This leads to the following result:

Theorem 4. *There exists an FPTAS for* SCP *in DAGs.*

Proof. Given $\varepsilon > 0$, run the dynamic programming algorithm from Sect. 4.3 $|\mathcal{V}| \cdot |V|$ times, once for each possible value of $c_{\max}$, and choose the best placement. According to Lemma 1 the best placement is $(1 + \varepsilon)$-approximate. The running time is polynomial in the input size and in $1/\varepsilon$. $\qquad\square$

5   General Networks

In this section we consider SCP when the physical network is an undirected graph. Recall that even the feasibility version of SCP is NP-hard in general networks, and therefore we focus on a special case. A vertex $v \in V$ is called *neighborly* if it has more than two neighbors. First, we assume that there exists an optimal placement whose physical path consists of at most k neighborly vertices. For example, this can be assumed if all simple paths from s to t contain at most k neighborly vertices. In this case we present a randomized algorithm that computes an $(1 + \varepsilon)$-approximate placement with high probability whose running time is $O(k! \cdot \text{poly}(n))$. Our second algorithm works under the stronger assumption that there are k neighborly vertices in the network. In this case, we present a deterministic algorithm whose running time is $O(k! \cdot \text{poly}(n))$.

Our randomized algorithm consists of $k!$ iterations of the following two phases: an orientation phase that applies a random orientation to the physical network, and an execution of the FPTAS for DAGs given in Theorem 4. The algorithm finds a $(1 - \varepsilon)$-approximate placement with probability $(1 - 1/e)$, and the running time of the algorithm is $O((n + t(n))k!)$, where $t(n)$ is the time it takes to compute a placement when the physical network is a DAG. As usual, one may amplify the probability of success using repetition.

The orientation phase is done as follows: let N be the set of neighborly vertices, and let $\pi : N \to \{1, \ldots, |N|\}$ be a random permutation. We direct the edges according to π. Observe that each edge $e \in E$ is found on a simple path between two neighborly nodes v_e and v'_e, in which all internal vertices are in $V \setminus N$. The edge e is directed towards v_e, if $\pi(v_e) > \pi(v'_e)$, and otherwise it is directed towards v'_e. Observe that when this process terminates we get a DAG, denoted by G_π, where there is a topological order than is consistent which π. Figure 5 depicts the orientation stage. A naïve implementation of this phase would run in $O(|V||E|)$ time.

Upon completing the orientation, we use our previous algorithm to find a placement. We repeat this process $k!$ times, each time with a new independent random permutation, and keep the best embedding so far.

Theorem 5. *There exists an algorithm, that given an* SCP *instance with an optimal solution that contains at most k neighborly vertices, finds a $(1 + \varepsilon)$-approximate placement with high probability, whose running time is $O(k! \cdot \text{poly}(n))$.*

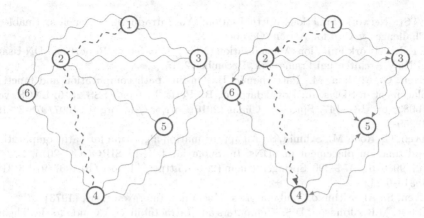

Fig. 5. Orientation phase: on the left is the physical network, where only the heavy nodes are drawn. The numbers represent the permutation and the dashed path represents a path of an optimal placement. The orientated physical network is on the right. It is enough that the internal ordering of the nodes on the dashed path is "correct" to ensure that the survival of the optimal placement.

Proof. Consider an optimal placement with at most k neighborly vertices, and let π' be the permutation it induces on the neighborly vertices in the physical path. If the random permutation π orders the neighborly vertices correctly, i.e., if $\pi(v) = \pi'(v)$, for every $v \in N$, then the optimal placement is feasible with respect to G_π. Hence, the FPTAS will find a $(1+\varepsilon)$-approximate placement in G_π. This placement is feasible, and therefore $(1+\varepsilon)$-approximate placement, with respect to G. The permutation π agrees with π' with probability $1/k!$, and the probability that π agrees with π' in $k!$ iterations is $1-(1-1/k!)^{k!} \geq 1-e^{-1}$. One may amplify the success probability to $1 - \frac{1}{n}$ using $\log n$ repetitions. $\square$

In the case where $|N| = k$ we can simply examine the $k!$ permutations of the neighborly vertices. For each possible permutation the algorithm applies the graph orientation procedure as described above, and then executes the FPTAS for DAGs given in Sect. 4.4. Thus the running time of the algorithm is $O(k! \cdot \text{poly}(n))$. We note that permutation enumeration for k numbers takes $O(k!)$ time (see e.g., [7,14]).

Theorem 6. *There exists an $O(k! \cdot \text{poly}(n))$-time algorithm, that given an SCP instance with k neighborly vertices, finds a $(1+\varepsilon)$-approximate placement.*

Acknowledgement. We thank Reuven Bar-Yehuda for helpful discussions.

References

1. Brown, G.: Service Chaining in Carrier Networks. Heavy Reading (2015)
2. Cohen, R., Lewin-Eytan, L., Naor, J., Raz, D.: Near optimal placement of virtual network functions. In: 34th IEEE Conference on Computer Communications, pp. 1346–1354 (2015)

3. ETSI: Network Functions Virtualisation: An Introduction, Benefits, Enablers, Challenges & Call for Action, October 2012
4. ETSI: Network Functions Virtualisation (NFV); Ecosystem; Report on SDN Usage in NFV Architectual Framework, December 2015
5. Even, G., Medina, M., Patt-Shamir, B.: On-line path computation and function placement in SDNs. In: Bonakdarpour, B., Petit, F. (eds.) SSS 2016. LNCS, vol. 10083, pp. 131–147. Springer, Cham (2016). https://doi.org/10.1007/978-3-319-49259-9_11
6. Even, G., Rost, M., Schmid, S.: An approximation algorithm for path computation and function placement in SDNs. In: Suomela, J. (ed.) SIROCCO 2016. LNCS, vol. 9988, pp. 374–390. Springer, Cham (2016). https://doi.org/10.1007/978-3-319-48314-6_24
7. Even, S.: Algorithmic Combinatorics. Macmillan Inc., New York (1973)
8. Garey, M.R., Johnson, D.S.: Computers and Intractability: A Guide to the Theory of NP-Completeness. W.H. Freeman and Company, San Francisco (1979)
9. Gember-Jacobson, A., Viswanathan, R., Prakash, C., Grandl, R., Khalid, J., Das, S., Akella, A.: OpenNF: enabling innovation in network function control. In: ACM SIGCOMM, pp. 163–174 (2014)
10. Hartert, R., Vissicchio, S., Schaus, P., Bonaventure, O., Filsfils, C., Telkamp, T., François, P.: A declarative and expressive approach to control forwarding paths in carrier-grade networks. In: ACM SIGCOMM, pp. 15–28 (2015)
11. Kreutz, D., Ramos, F.M.V., Veríssimo, P.J.E., Rothenberg, C.E., Azodolmolky, S., Uhlig, S.: Software-defined networking: a comprehensive survey. Proc. IEEE **103**(1), 14–76 (2015)
12. Lukovszki, T., Schmid, S.: Online admission control and embedding of service chains. In: Scheideler, C. (ed.) Structural Information and Communication Complexity. LNCS, vol. 9439, pp. 104–118. Springer, Cham (2015). https://doi.org/10.1007/978-3-319-25258-2_8
13. Rost, M., Schmid, S.: Service chain and virtual network embeddings: Approximations using randomized rounding. Technical report abs/1604.02180, CoRR (2016)
14. Sedgewick, R.: Permutation generation methods. ACM Comput. Surv. **9**(2), 137–164 (1977)
15. Soulé, R., Basu, S., Marandi, P.J., Pedone, F., Kleinberg, R.D., Sirer, E.G., Foster, N.: Merlin: a language for provisioning network resources. In: 10th ACM International on Conference on emerging Networking Experiments and Technologies, pp. 213–226 (2014)

Tight Approximability of the Server Allocation Problem for Real-Time Applications

Takehiro Ito[1], Naonori Kakimura[2(✉)], Naoyuki Kamiyama[3],
Yusuke Kobayashi[4], Yoshio Okamoto[5], and Taichi Shiitada[5]

[1] Tohoku University, Sendai, Japan
takehiro@ecei.tohoku.ac.jp
[2] Keio University, Yokohama, Japan
kakimura@math.keio.ac.jp
[3] Kyushu University, Fukuoka, Japan
kamiyama@imi.kyushu-u.ac.jp
[4] University of Tsukuba, Tsukuba, Japan
kobayashi@sk.tsukuba.ac.jp
[5] University of Electro-Communications, Chofu, Japan
okamotoy@uec.ac.jp, shiitada@gmail.com

Abstract. The server allocation problem is a facility location problem for a distributed processing scheme on a real-time network. In this problem, we are given a set of users and a set of servers. Then, we consider the following communication process between users and servers. First a user sends his/her request to the nearest server. After receiving all the requests from users, the servers share the requests. A user will then receive the data processed from the nearest server. The goal of this problem is to choose a subset of servers so that the total delay of the above process is minimized. In this paper, we prove the following approximability and inapproximability results. We first show that the server allocation problem has no polynomial-time approximation algorithm unless $\mathbf{P} = \mathbf{NP}$. However, assuming that the delays satisfy the triangle inequality, we design a polynomial-time $\frac{3}{2}$-approximation algorithm. When we assume the triangle inequality only among servers, we propose a polynomial-time 2-approximation algorithm. Both of the algorithms are tight in the sense that we cannot obtain better polynomial-time approximation algorithms unless $\mathbf{P} = \mathbf{NP}$. Furthermore, we evaluate the practical performance of

T. Ito—Supported by JST CREST Grant Number JPMJCR1402, Japan, and JSPS KAKENHI Grant Number JP16K00004.
N. Kakimura—Supported by JST ERATO Grant Number JPMJER1201, Japan, and by JSPS KAKENHI Grant Number JP17K00028.
N. Kamiyama—Supported by JST PRESTO Grant Number JPMJPR14E1, Japan.
Y. Kobayashi—Supported by JST ERATO Grant Number JPMJER1201, Japan, and by JSPS KAKENHI Grant Numbers JP16K16010 and JP16H03118.
Y. Okamoto—Supported by Kayamori Foundation of Informational Science Advancement, JST CREST Grant Number JPMJCR1402, Japan, and JSPS KAKENHI Grant Numbers JP24106005, JP24700008, JP24220003, JP15K00009.

© Springer International Publishing AG, part of Springer Nature 2018
D. Alistarh et al. (Eds.): ALGOCLOUD 2017, LNCS 10739, pp. 41–55, 2018.
https://doi.org/10.1007/978-3-319-74875-7_4

our algorithms through computational experiments, and show that our algorithms scale better and produce comparable solutions than the previously proposed method based on integer linear programming.

1   Introduction

Nowadays, various kinds of services such as video conferencing and document sharing are offered through a network, which are getting indispensable in our daily life. They provide an online platform for users to exchange their information with each other. Some of online services, such as online games and ticket reservation systems, are accessed by many users at the same time, which requires a real-time process.

An online service is usually maintained on one central application server to keep its integrity and maintainability. It brings in a heavy computing load on the server. Not only that, more crucially, it will cause different latencies on users. Since it takes more time for a user further from the server to send her request, a request of a further user arrives later even when she sent the request earlier. Since the ordering of requests is essential in real-time processing such as online games, we have to wait for a request from the farthest user to arrive before processing any request. After receiving all the requests from users, the server can process them in the ordering of requested time, which is estimated from the arrival times. Thus latencies of a server response occur due to the order adjustment of requests, which makes a real-time interaction inefficient.

Recently, Kawabata *et al.* [11] proposed a distributed processing scheme for a real-time network to reduce the latency. In this scheme, instead of having one central server, we set up a set of application servers that provide the same service. It allows every user to access to one of the servers quickly. However, since each user sends their request to different servers, each server needs to multi-cast the received requests to the other servers to synchronize them. It enables each server to process all the requests in the right order of the requested times. A user will then receive the data processed successfully from the nearest server. Thus, the proposed scheme has an additional step to synchronize the data among the servers, but, since each user has a close server, we can reduce the latency caused by the order adjustment.

Let us define the problem more formally. We are given a set U of users and a set S of possible places that we can allocate our application servers. For each pair $i, j \in S \cup U$, we denote by $d(i, j)$ the delay of the communication between i and j. We assume that $d(i, j) \geq 0$, $d(i, j) = d(j, i)$, and $d(i, i) = 0$ for any i, j. Suppose that we choose a set X of servers in S. Then each user u accesses to her nearest server in X, which is denoted by $\rho(u, X)$. That is, $\rho(u, X) \in \mathrm{argmin}\{d(u, s) \mid s \in X\}$. As we have to wait for the last request, it takes $\max_{u \in U} d(u, \rho(u, X))$ time to receive all the requests. After that, the servers in X communicate with each other for synchronization, which takes $\max_{s, t \in X} d(s, t)$ time. Finally, the processed data is returned to every user in $\max_{u \in U} d(u, \rho(u, X))$ time. Thus the total delay of the whole process, denoted by $\mathsf{delay}(X)$, is defined to be

$$\text{delay}(X) := \max_{u \in U} 2 \cdot d(u, \rho(u, X)) + \max_{s,t \in X} d(s,t).$$

We want to find a set X of servers that minimizes the total delay. The problem is called *the server allocation problem for real-time applications*, but *the server allocation problem* for short.

Kawabata et al. [11] introduced the server allocation problem and formulated it as an integer linear programming problem. They also performed numerical simulations to measure the performance of their distributed server allocation scheme. Furthermore, Ba et al. [2] later proved that, when each application server has a limited capacity of users, the problem is **NP**-hard by reduction from 3SAT. However, it was not known whether the server allocation problem is polynomial-time solvable or not.

In this paper, we show that the server allocation problem is **NP**-hard. In fact, it is impossible to get any polynomial-time approximation algorithm unless **P = NP**. On the positive side, assuming that the delay d is metric, we design a polynomial-time constant-factor approximation algorithm for the problem. For a subset Z in $S \cup U$, we say that the delay d is *metric in* Z if for every triple of elements i_1, i_2, i_3 in Z, d obeys the *triangle inequality*, i.e., $d(i_1, i_2) + d(i_2, i_3) \geq d(i_1, i_3)$. We consider the following three cases (see also Table 1).

Case 1: Metric in $S \cup U$. We design a polynomial-time $\frac{3}{2}$-approximation algorithm when d is metric in $S \cup U$. Moreover, we prove that, for any constant $r < \frac{3}{2}$, there exists no polynomial-time r-approximation algorithm unless **P = NP**.

Case 2: Metric in S. We design a polynomial-time 2-approximation algorithm when d is metric in S. Moreover, we show that, for any constant $r < 2$, there exists no polynomial-time r-approximation algorithm unless **P = NP**.

Case 3: General Case. We show that there exists no polynomial-time r-approximation algorithm for any r unless **P = NP**.

The inapproximability can be strengthened if we assume the *exponential-time hypothesis*, which basically states that we cannot solve 3SAT in $2^{o(n)}$ time [9,10]. Namely, we show that those inapproximability results follow even if we relax the running times to $2^{o(n)}$, where n is the number of servers, assuming the exponential-time hypothesis holds.

Table 1. Summary of our result

	Condition	Approximability	Inapproximability
Case 1	Metric in $S \cup U$	1.5	$1.5 - \varepsilon$
Case 2	Metric in S	2	$2 - \varepsilon$
Case 3	General	—	∞

Let us conclude this section with describing related work. The server allocation problem was proposed by Kawabata et al. [11]. They introduced a more general problem, in which users may give up sending a request if they are far from the chosen servers. Our problem is a special case of their problem when all users have to access to one of the chosen servers.

Our problem is related to the *k-center problem*. In the k-center problem, we are given a set of users and a set of facilities. The goal is to find a set X of at most k facilities that minimizes the maximum distance from the users to X. The k-center problem is **NP**-hard, but has polynomial-time 2-approximation algorithms [7,8] if the set of users and facilities satisfies the triangle inequality. Furthermore, for any constant $r <\cdot 2$, there exists no polynomial-time r-approximation algorithm unless $\mathbf{P} = \mathbf{NP}$ [7,8]. Our problem differs in that we have no constraints on the number of opened facilities. Instead, the more we open facilities, the more communication cost between facilities is required, which is taken into the objective value. The *k-median problem* is also related to our problem. In this problem, we are given a set of users and a set of facilities. Then, the goal of this problem is to find a set X of at most k facilities that minimizes the sum of distances from the users to X. If the set of users and facilities satisfies the triangle inequality, then the current best polynomial-time approximation ratio for this problem is $2.611 + \varepsilon$ due to Byrka *et al.* [3]. Our problem differs in that our goal is to minimize the maximum delay.

The *hub location problem* is the problem of locating hubs and allocate non-hub nodes to decrease the transportation cost between a pair of nodes. The problem has been studied in transportation and telecommunication systems (see e.g., [1,6] for recent surveys). The *k-hub center problem* is a variant of the hub location problem in which we locate k hubs so that the maximum distance between any pair of nodes is minimized [4,12]. Recently, Chen *et al.* [5] proposed polynomial-time constant-factor approximation algorithms for the k-hub center problem.

2   Approximation Algorithms

In this section, we design approximation algorithms for the server allocation problem.

Let X^* be an optimal solution. We denote

$$k^* = \max_{u \in U} d(u, \rho(u, X^*)) \quad \text{and} \quad \ell^* = \max_{s,t \in X^*} d(s,t).$$

Then, $\mathsf{delay}(X^*) = 2k^* + \ell^*$.

Also, define $D_1 := \{d(s,u) \mid (s,u) \in S \times U\}$ and $D_2 := \{d(s,t) \mid (s,t) \in S^2\}$. It should be noted that $|D_1| = O(|S||U|)$, $|D_2| = O(|S|^2)$, $k^* \in D_1$ and $\ell^* \in D_2$.

2.1   Case 1: Metric in $S \cup U$

In this section, we assume that the delay d is metric in $S \cup U$, that is, for every triple of elements i_1, i_2, i_3 in $S \cup U$, the triangle inequality holds, i.e., $d(i_1, i_2) + d(i_2, i_3) \geq d(i_1, i_3)$. The idea of our algorithm is to perform two different algorithms, and then the better one of the two results gives a $\frac{3}{2}$-approximate solution.

The first algorithm simply finds a server s that minimizes $\mathsf{delay}(\{s\})$.

Lemma 1. *Let $Z_1 := \{s'\}$, where $s' \in \text{argmin}\{\text{delay}(\{s\}) \mid s \in S\}$. Then,* $\text{delay}(Z_1) \leq 2k^* + 2\ell^*$.

Proof. Let s be a server in the optimal solution X^*. Then, by the definition of Z_1, we have

$$\text{delay}(Z_1) \leq \text{delay}(\{s\}) = \max_{u \in U} 2 \cdot d(s, u).$$

Since d is metric in $S \cup U$, it follows that

$$d(s, u) = d(u, s) \leq d(u, \rho(u, X^*)) + d(\rho(u, X^*), s) \leq k^* + \ell^*,$$

where the last inequality follows from the fact that $d(u, \rho(u, X^*)) \leq k^*$ and $d(\rho(u, X^*), s) \leq \ell^*$ as $\rho(u, X^*) \in X^*$. Therefore, combining the above two inequalities, we obtain $\text{delay}(Z_1) \leq 2k^* + 2\ell^*$. $\qquad\square$

For the second algorithm, we first consider

$$X_u := \{s \in S \mid d(s, u) \leq k^*, \ d(s, u') \leq k^* + \ell^* \ (\forall u' \in U)\},$$

for each user u in U, and define $Z_2^* := \bigcup_{u \in U} X_u$. Intuitively, for each user u in U, the set X_u is a set of servers that are close to u and the other users simultaneously.

Lemma 2. *Suppose that Z_2^* is defined as above. Then,* $\text{delay}(Z_2^*) \leq 4k^* + \ell^*$.

Proof. Let s be a server in Z_2^*. By definition of Z_2^*, there exists a user u_s such that $d(s, u_s) \leq k^*$. Moreover, for any other server t in Z_2^*, it holds that $d(u_s, t) \leq k^* + \ell^*$. Since d is metric in $S \cup U$, we have

$$d(s, t) \leq d(s, u_s) + d(u_s, t) \leq k^* + (k^* + \ell^*) = 2k^* + \ell^*.$$

Hence it holds that $\max_{s,t \in Z_2^*} d(s, t) \leq 2k^* + \ell^*$.

Let u be a user in U. Define $s := \rho(u, X^*)$. Then, $d(u, s) \leq k^*$ holds by definition. Since d is metric in $S \cup U$, for every user u' in U, $d(u', s) \leq d(u', \rho(u', X^*)) + d(\rho(u', X^*), s)$. By definition of k^*, we have $d(u', \rho(u', X^*)) \leq k^*$. Moreover, since $\rho(u', X^*) \in X^*$, we see $d(s, \rho(u', X^*)) \leq \ell^*$. Therefore, we obtain $d(u', s) \leq k^* + \ell^*$. This, together with that $d(u, s) \leq k^*$, implies $s \in X_u$, and thus $s \in Z_2^*$. Hence, since $d(u, \rho(u, Z_2^*)) \leq d(u, s) \leq k^*$, we obtain $\max_{u \in U} 2d(u, \rho(u, Z_2^*)) \leq 2k^*$.

Therefore, it follows that

$$\text{delay}(Z_2^*) = \max_{u \in U} 2 \cdot d(u, \rho(u, Z_2^*)) + \max_{s,t \in Z_2^*} d(s, t) \leq 2k^* + (2k^* + \ell^*) = 4k^* + \ell^*.$$

$\qquad\square$

To find the set Z_2^*, we need to know k^* and ℓ^* in advance. We search for the exact values of k^* and ℓ^* by enumerating all the elements in D_1 and D_2, respectively. For each candidate, we construct the corresponding set Z_2^*. Since $k^* \in D_1$ and $\ell^* \in D_2$, at least one candidate returns the set X with $\text{delay}(X) \leq 4k^* + \ell^*$.

Our algorithm for Case 1 is described as Algorithm 1. We now show the approximation factor of $\frac{3}{2}$.

Algorithm 1. Algorithm for Case 1 (Metric in $S \cup U$)

Step 1. Compute D_1 and D_2.

Step 2. Find $Z_1 := \{s'\}$, where $s' \in \mathrm{argmin}\{\mathsf{delay}(\{s\}) \mid s \in S\}$.

Step 3. For each element $k \in D_1$ and each element $\ell \in D_2$, define

$$X_u^{k,\ell} := \{s \in S \mid d(s,u) \le k,\ d(s,u') \le k + \ell \ (\forall u' \in U)\} \text{ for each user } u \text{ in } U,$$

and construct $X^{k,\ell} := \bigcup_{u \in U} X_u^{k,\ell}$.

Step 4. Find $\min\{\mathsf{delay}(X^{k,\ell}) \mid (k,\ell) \in D_1 \times D_2\}$, and let Z_2 be the set $X^{k,\ell}$ attaining the minimum.

Step 5. If $\mathsf{delay}(Z_1) < \mathsf{delay}(Z_2)$, then return $Z := Z_1$, and otherwise, return $Z := Z_2$.

Theorem 1. *Let Z be the output of Algorithm 1. Then, $\mathsf{delay}(Z) \le \frac{3}{2} \cdot \mathsf{delay}(X^*)$.*

Proof. Since $\mathsf{delay}(Z_2) \le \mathsf{delay}(Z_2^*)$, Lemmas 1 and 2 imply that

$$\mathsf{delay}(Z) \le \min\{\mathsf{delay}(Z_1), \mathsf{delay}(Z_2^*)\} \le \frac{1}{2}(\mathsf{delay}(Z_1) + \mathsf{delay}(Z_2^*))$$

$$\le \frac{1}{2}((2k^* + 2\ell^*) + (4k^* + \ell^*)) = \frac{3}{2}(2k^* + \ell^*) = \frac{3}{2} \cdot \mathsf{delay}(X^*).$$

□

Since $|D_1|$ and $|D_2|$ are polynomially bounded by $|S|$ and $|U|$, Algorithm 1 runs in polynomial time. More specifically, Step 1 takes $O(|S|^2 + |S||U|)$ time. In Step 2, $\mathsf{delay}(\{s\})$ can be computed in $O(|U|)$ time for each $s \in S$, and thus Step 2 takes $O(|S||U|)$ time. In Step 3, $X_u^{k,\ell}$ can be found in $O(|S||U|)$ time for each $k \in D_1$, $\ell \in D_2$, and $u \in U$, and thus Step 3 takes $O(|S||U|) \times O(|D_1||D_2||U|) = O(|S|^4|U|^3)$ time. In Step 4, $\mathsf{delay}(X^{k,\ell})$ can be computed in $O(|S|^2 + |S||U|)$ time for each $(k,\ell) \in D_1 \times D_2$, and thus Step 4 takes $O(|S|^2 + |S||U|) \times O(|D_1||D_2|) = O(|S|^5|U| + |S|^4|U|^2)$ time. Step 5 takes $O(1)$ time as we already know $\mathsf{delay}(Z_1)$ and $\mathsf{delay}(Z_2)$. Hence, in total, Algorithm 1 runs in $O(|S|^5|U| + |S|^4|U|^3)$ time. In Sect. 4.1, we improve to $O(|S|^2(|S| + |U|))$.

2.2   Case 2: Metric in S

In this section, we assume that the delay d is metric in S (not necessarily in $S \cup U$). Similarly to Case 1, we use the values k^* and ℓ^* of the optimal solution X^*.

Theorem 2. *Define $X_s := \{t \in S \mid d(s,t) \le \ell^*\}$ for each server s in S, and define $Z_3^* := X_{s'}$ where $s' \in \mathrm{argmin}\{\mathsf{delay}(X_s) \mid s \in S\}$. Then, it holds that $\mathsf{delay}(Z_3^*) \le 2k^* + 2\ell^* \le 2 \cdot \mathsf{delay}(X^*)$.*

Proof. Let s be a server in the optimal solution X^*. Since $\mathsf{delay}(Z_3^*) \le \mathsf{delay}(X_s)$, it suffices to show $\mathsf{delay}(X_s) \le 2k^* + 2\ell^*$.

Algorithm 2. Algorithm for Case 2 (Metric in S)

Step 1. Compute D_2.

Step 2. For each element $\ell \in D_2$, define

$$X_s^\ell := \{t \in S \mid d(s,t) \le \ell\}$$

for each server $s \in S$. Find $\min\{\mathsf{delay}(X_s^\ell) \mid s \in S\}$, and let X^ℓ be the set attaining the minimum.

Step 3. Find $\min\{\mathsf{delay}(X^\ell) \mid \ell \in D_2\}$, and return the set attaining the minimum as Z_3.

Since $s \in X^*$, we have $d(s,t) \le \ell^*$ for every server t in X^*. This implies that $X^* \subseteq X_s$. Hence, for every user u in U, $d(u, \rho(u, X_s)) \le d(u, \rho(u, X^*)) \le k^*$. Furthermore, since d is metric in S, for every pair of servers t, t' in X_s,

$$d(t,t') \le d(t,s) + d(s,t') \le \ell^* + \ell^* = 2\ell^*,$$

where the last inequality follows since $s, t, t' \in X_s$. Therefore,

$$\mathsf{delay}(X_s) = \max_{u \in U} 2 \cdot d(u, \rho(u, X_s)) + \max_{t,t' \in X_s} d(t,t') \le 2k^* + 2\ell^*. \qquad \square$$

To find the set Z_3^*, we enumerate all the elements in D_2 to search for the exact value of ℓ^*. Our algorithm for Case 2 is described as Algorithm 2. It follows from Theorem 2 that the output of Algorithm 2 is a 2-approximate solution. Moreover, the algorithm runs in polynomial time, as $|D_2| = O(|S|^2)$. More specifically, Step 1 takes $O(|S|^2)$ time. In Step 2, the determination of X_s^ℓ takes $O(|S|)$ time and the computation of $\mathsf{delay}(X_s^\ell)$ takes $O(|S||U|)$ time for each $\ell \in D_2$ and $s \in S$, and thus Step 2 takes $O(|D_2||S|^2|U|) = O(|S|^4|U|)$ time. Step 3 takes $O(|D_2|) = O(|S|^2)$ time. Hence, the running time of Algorithm 2 is bounded by $O(|S|^4|U|)$.

In Sect. 4.1, we improve the running time to $\tilde{O}(|S|(|S| + |U|))$, where the $\tilde{O}$-notation suppresses the polylogarithmic factor. Note that $\Omega(|S|(|S| + |U|))$ time is needed to read off the entire delay d.

3 Hardness of Approximation

In this section, we prove the inapproximability of the server allocation problem by reduction from the well-known **NP**-complete problem 3SAT.

Assume that we are given an instance I of 3SAT with variables $x_1, x_2, \ldots, x_n$ and clauses $C_1, C_2, \ldots, C_m$, where $m = O(n)$. From the instance, we construct an instance of the server allocation problem I' as follows. We define $S := \{x_1, x_2, \ldots, x_n, \overline{x_1}, \overline{x_2}, \ldots, \overline{x_n}\}$ and $U := \{C_1, C_2, \ldots, C_m\}$. For each pair of servers s, s' in S, we define

$$d(s, s') := \begin{cases} \alpha & \text{if } \{s, s'\} \neq \{x_i, \overline{x_i}\} \text{ for any } i, \\ \beta & \text{otherwise}, \end{cases}$$

where $\alpha < \beta$. Furthermore, for each server s in S and each user u in U, we define

$$d(s, u) := \begin{cases} \gamma & \text{if } s \text{ is a literal in } u, \\ \delta & \text{otherwise}, \end{cases}$$

where $\gamma < \delta$. Then the following lemma holds.

Lemma 3. *A 3SAT instance I is a yes-instance if and only if the optimal value of the instance I' is at most $2\gamma + \alpha$. On the other hand, I is a no-instance if and only if the optimal value of I' is at least $\min\{2\gamma + \beta, 2\delta + \alpha\}$.*

Proof. Assume that I is a yes-instance. Then, there exists a satisfying assignment $a\colon \{x_1, x_2, \ldots, x_n\} \to \{\text{true}, \text{false}\}$ for I. We define $X := \{x_i \mid a(x_i) = \text{true}\} \cup \{\overline{x_i} \mid a(x_i) = \text{false}\}$. Since the set X has either x_i or $\overline{x_i}$ for every i, it holds that $\max_{s, s' \in X} d(s, s') \leq \alpha$. Moreover, by definition of d, we have $\max_{u \in U} d(u, \rho(u, X)) \leq \gamma$. Thus the optimal value of I' is at most $2\gamma + \alpha$. Conversely, if the optimal value of I' is at most $2\gamma + \alpha$, then there exists a set X of servers such that $\max_{s, s' \in X} d(s, s') \leq \alpha$ and $\max_{u \in U} d(u, \rho(u, X)) \leq \gamma$. Hence, X contains either x_i or $\overline{x_i}$ for every i. We set x_i to be true if $x_i \in X$ and false otherwise. This is a satisfying assignment since each clause C_j has a server within the distance γ.

The second statement follows since the objective value of I' can only be one of the following four values: $2\gamma + \alpha$, $2\delta + \alpha$, $2\gamma + \beta$, and $2\delta + \beta$. $\qquad\square$

The above lemma immediately implies the **NP**-hardness for the server allocation problem. Moreover, the reduction introduces the gap of at least $\min\{2\gamma + \beta, 2\delta + \alpha\}/(2\gamma + \alpha)$. By setting the appropriate values for those four parameters, we obtain the tight inapproximability results.

Theorem 3. *(1) For any r, there is no polynomial-time r-approximation algorithm for the server allocation problem unless $\mathbf{P} = \mathbf{NP}$.*

(2) Suppose that d is metric in $S \cup U$. Then, for any constant $r < \frac{3}{2}$, there exists no polynomial-time r-approximation algorithm for the server allocation problem unless $\mathbf{P} = \mathbf{NP}$.

(3) Suppose that d is metric in S. Then, for any constant $r < 2$, there exists no polynomial-time r-approximation algorithm for the server allocation problem unless $\mathbf{P} = \mathbf{NP}$.

Proof. (1) Set $\alpha := 1$, $\beta := \infty$, $\gamma := 1$, and $\delta := \infty$ in the definition of d. It follows from Lemma 3 that we cannot distinguish $2\gamma + \alpha$ and $\min\{2\gamma + \beta, 2\delta + \alpha\}$ in polynomial time unless $\mathbf{P} = \mathbf{NP}$. The ratio of the two values is

$$\frac{\min\{2\gamma + \beta, 2\delta + \alpha\}}{2\gamma + \alpha} = \frac{\min\{\infty, \infty\}}{3} = \infty.$$

Thus we cannot have a polynomial-time r-approximation algorithm for any r.

(2) Set $\alpha := 2$, $\beta := 4$, $\gamma := 1$, and $\delta := 3$ in the definition of d. Note that d is metric in $S \cup U$, as we may assume that a 3SAT instance I does not contain a clause having both x_i and $\overline{x_i}$ for some i. It follows from Lemma 3 that we cannot distinguish $2\gamma + \alpha$ and $\min\{2\gamma + \beta, 2\delta + \alpha\}$ in polynomial time unless $\mathbf{P} = \mathbf{NP}$. The ratio of the two values is

$$\frac{\min\{2\gamma + \beta, 2\delta + d\}}{2\gamma + \alpha} = \frac{\min\{6, 8\}}{4} = \frac{3}{2}.$$

(3) Set $\alpha := 1$, $\beta := 2$, $\gamma := \varepsilon$, and $\delta = \infty$ in the definition of d, where $\varepsilon > 0$ is an arbitrarily small constant. Then, d is metric in S. It follows from Lemma 3 that, unless $\mathbf{P} = \mathbf{NP}$, we cannot have a polynomial-time r-approximation algorithm, where r is a constant less than

$$\frac{\min\{2\gamma + \beta, 2\delta + \alpha\}}{2\gamma + \alpha} = \frac{\min\{2\varepsilon + 2, \infty\}}{2\varepsilon + 1} \to 2 \qquad (\text{as } \varepsilon \to 0). \qquad \square$$

In our reduction in Lemma 3, the number of servers is $2n$ and the number of users can be $m = O(n)$. Thus, we may conclude that if the server allocation problem can be solved in $2^{o(n)}$ time, where n is the number of servers, then the exponential-time hypothesis fails (see [9,10] for the precise definition). The same consequence can be obtained for the inapproximability, which we summarize in the following theorem.

Theorem 4. *If any of the following (1)–(3) holds, then the exponential-time hypothesis fails. Here, n is the number of servers.*
(1) *The server allocation problem can be approximated by any r in $2^{o(n)}$ time.*
(2) *The server allocation problem can be approximated by any constant $r < 2$ in $2^{o(n)}$ time when d is metric in S.*
(3) *The server allocation problem can be approximated by any constant $r < 3/2$ in $2^{o(n)}$ time when d is metric in $S \cup U$.* $\qquad \square$

Note that the existence of an exact algorithm running in $O(2^{|S|}\text{poly}(|S|, |U|))$ time is clear: We just need to go through all the possible candidates $X \subseteq S$, compute their delay, and take the minimum.

4   Experiments

Since our algorithms are combinatorial and simple, we conducted computational experiments to evaluate the practical performance of our algorithms. The performance is measured in terms of execution time and the objective value of an obtained solution. The baseline is an integer linear programming (ILP) formulation given by Kawabata et al. [11]. The ILP instances were solved by IBM ILOG CPLEX 12.7.0.0. The implementation was done by C++ (gcc 5.4.0), and all the programs were run on a machine with the following specification: OS Ubuntu 16.04.1 LTS 64 bit, Memory 3.8 GiB, Processor Intel Core i3-2120 CPU 3.30 GHz × 4, HDD 488.0 GB.

Algorithm 3. Accelerated Algorithm 1

Step 1-1. Compute D_1 and D_2 and sort D_1 and D_2 in descending order.

Step 1-2. For each server s in S, find

$$u_s^{\min} \in \mathrm{argmin}\{d(s,u) \mid u \in U\} \quad \text{and} \quad u_s^{\max} \in \mathrm{argmax}\{d(s,u) \mid u \in U\},$$

and sort S by the descending order of $d(s, u_s^{\min})$.

Step 2. Find $Z_1 := \{s'\}$, where $s' \in \mathrm{argmin}\{\mathsf{delay}(\{s\}) \mid s \in S\}$.

Step 3. For each element $k \in D_1$ in descending order, do the following.

Step 3-1. Determine the set

$$Y^k := \{s \in S \mid d(s, u_s^{\min}) \leq k\}$$

by linear search over S, and sort Y^k in the descending order of $d(s, u_s^{\max})$. Let $Y^k = \{s_1, s_2, \ldots, s_{t(k)}\}$ in that order.

Step 3-2. For each $i \in \{1, 2, \ldots, t(k)\}$, determine

$$\ell_i := \min\{\ell \in D_2 \mid d(s_i, u_{s_i}^{\max}) \leq \ell + k\}$$

by binary search over D_2, and determine $X^{k,\ell_i} := \{s \in Y^k \mid d(s, u_s^{\max}) \leq k + \ell_i\}$ by linear search over Y^k.

Step 4. Find $\min\{\mathsf{delay}(X^{k,\ell_i}) \mid k \in D_1, i \in \{1, 2, \ldots, t(k)\}\}$, and let Z_2 be the set X^{k,ℓ_i} attaining the minimum.

Step 5. If $\mathsf{delay}(Z_1) < \mathsf{delay}(Z_2)$, then return $Z := Z_1$, and otherwise, return $Z := Z_2$.

4.1 Acceleration of the Proposed Algorithms

A naive implementation of our algorithms is quite slow. As described before, Algorithm 1 for Case 1 runs in $O(|S|^5|U| + |S|^4|U|^3)$ time, and Algorithm 2 for Case 2 runs in $O(|S|^4|U|)$ time. To speed-up the algorithms without losing approximability, we inserted preprocessing steps to sort the sets D_1 and D_2.

For Case 1, an easy but useful observation is the following. For a server $s \in S$, let

$$u_s^{\min} \in \mathrm{argmin}\{d(s,u) \mid u \in U\} \quad \text{and} \quad u_s^{\max} \in \mathrm{argmax}\{d(s,u) \mid u \in U\}.$$

Namely, $u_s^{\min}$ is the closest user to s and $u_s^{\max}$ is the furthest user from s. Then, $s \in X^{k,\ell}$ if and only if $d(s, u_s^{\min}) \leq k$ and $d(s, u_s^{\max}) \leq k + \ell$. This observation makes the determination of $X^{k,\ell}$ easier. Furthermore, the number of different $X^{k,\ell}$ is small: it can be bounded by $O(|D_1||D_2|)$, but can be reduced to $O(|S|^2)$. Those observations result in speed-up of the algorithm.

The modified algorithm for Case 1 is described as Algorithm 3. We will see that it runs in $O(|S|^2(|S| + |U|))$ time. In fact, Step 1-1 takes $O(|S|(|S| + |U|) + |D_1|\log|D_1| + |D_2|\log|D_2|) = O(|S||U|(\log|S| + \log|U|) + |S|^2\log|S|)$ time. Step 1-2 takes $O(|S||U| + |S|\log|S|)$ time. Step 2 takes $O(|S||U|)$ time as Algorithm 1. In Step 3-1, the linear search takes $O(|D_1| + |S|)$ time, and the sorting takes $O(|S|\log|S|)$ time by insertion sort in total since we only have at most $|S|$ distinct sets as Y^k. Thus, Step 3-1 takes $O(|S|(|U| + \log|S|))$ time. Step 3-2 takes

---

Algorithm 4. Accelerated Algorithm 2

Step 1-1. Compute D_2 and sort D_2 in descending order.

Step 1-2. For each server s in S, define

$$D_{2,s} := \{d(s,t) \mid t \in S\},$$

and sort $D_{2,s}$ in decreasing order. Note that $D_2 = \bigcup_{s \in S} D_{2,s}$ as a set.

Step 2. For each server s in S and each element ℓ in $D_{2,s}$ in decreasing order, determine

$$X_s^\ell := \{t \in S \mid d(s,t) \le \ell\}$$

by linear search, and find $\min\{\mathsf{delay}(X_s^\ell) \mid s \in S\}$, and let X^ℓ be the set attaining the minimum.

Step 3. Find $\min\{\mathsf{delay}(X^\ell) \mid \ell \in D_2\}$, and return the set attaining the minimum as Z_3.

$O(|Y^k| \log |D_2|) = O(|S| \log |S|)$ time. Step 4 can be executed in $O(|S|^2(|S| + |U|))$ time since we only have at most $|S|$ distinct sets as Y^k, and for each distinct Y^k we scan the relevant entries of d (the number of relevant entries is $O(|S|(|S| + |U|))$). Step 5 takes $O(1)$ time. Therefore, in total, Algorithm 3 runs in $O(|S|^2(|S| + |U|))$ time.

For Case 2, we obtain a similar speed-up, as described in Algorithm 4. The algorithm runs in $O(|S|^2 \log |S| + |S||U|)$ time, which is $\tilde{O}(|S|(|S| + |U|))$. More specifically, Step 1-1 takes $O(|D_2| \log |D_2|) = O(|S|^2 \log |S|)$ time, and Step 1-2 takes $O(|S|^2 \log |S|)$ time in total. In Step 2, with the help of sorted orders, the computation takes $O(|S|(|S| + |U|))$ time. Step 3 takes $O(|D_2|) = O(|S|^2)$ time. Therefore, in total, Algorithm 4 runs in $O(|S|^2 \log |S| + |S||U|)$ time, which is $\tilde{O}(|S|(|S| + |U|))$.

Note that in Algorithms 3 and 4, we do not make use of the experiment setup where the instances are given on the Euclidean plane. They can run on any instances of the problem.

As a preliminary experiment showed, Algorithm 3 was still slow in practice. Thus, we consider a heuristic improvement of running time by losing the theoretical guarantee for the approximation ratio. The idea is a simple interleaving. After we construct the set D_1, we throw most of the elements in D_1 away, but we only keep at most $\sqrt{|D_1|}$ elements. Namely, we only keep every $\sqrt{|D_1|}$-th elements after the set is sorted. The speed-up is gained by replacing every occurrence of $|D_1|$ by $\sqrt{|D_1|}$, in the running time analysis after sorting D_1. Then, we conclude that the heuristic interleaving of Algorithm 3 results in an algorithm that yet runs in $O(|S|^2(|S| + |U|))$ time, but practically it is much faster than Algorithm 3 as we will see.

In the following experiments, algo1 refers to the accelerated version of Algorithm 1 (thus, Algorithm 3), algo2 refers to the accelerated version of Algorithm 2 (thus, Algorithm 4), algo1-red refers to the accelerated version of Algorithm 1 with the heuristic interleaving, and CPLEX refers to the ILP formulation solved by CPLEX 12.7.0.0.

4.2   Experiment 1: Following Kawabata *et al.* [11]

In Experiment 1, we follow the experiment setup by Kawabata *et al.* [11]. The users are distributed uniformly at random in the square $[0, 20] \times [0, 20]$. The number of servers is fixed to eight, and they are placed at the position of vertices of a regular octagon, which is centered at $(10, 10)$ and has diameter 10. The number of users ranges from 10 to 960 with increment of 50. For each number of users, 20 data are generated. Delay is simply defined by the Euclidean distance, and thus is metric in $S \cup U$. See Fig. 1 (left).

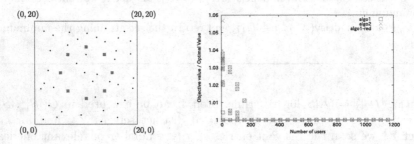

Fig. 1. (Left) The setup of Experiment 1. Red squares are servers, and black dots are users. (Right) Comparison of objective values in Experiment 1. (Color figure online)

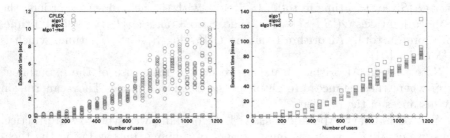

Fig. 2. (Left) Comparison of execution times in Experiment 1. (Right) The same plot as left, but the CPLEX data are dropped.

Figure 1 (right) shows the comparison of objective values. The vertical axis refers to the approximation ratio, where the optimal values are computed by CPLEX. When the number of users is small (up to 160), the proposed algorithms sometimes output suboptimal solutions, but when the number of users is larger, they always output optimal solutions. Even in the worst case, the approximation ratio is less than 1.06.

Figure 2 compares the execution times. As the left figure shows, the proposed algorithms run much faster than CPLEX. The right figure shows the comparison between the proposed algorithms, without CPLEX. We can see that algo1 is slower than algo2 and algo1-red that have the comparable performance.[1]

As a summary of Experiment 1, we conclude that the proposed algorithms scale better than CPLEX, and the solution quality is not as bad as the theoretical guarantees predict.

4.3   Experiment 2: With More Servers

In Experiment 2, we again follow the experiment setup by Kawabata *et al.* [11] as Experiment 1, but the number of servers will be changed. The servers and users are distributed uniformly at random in the square $[0, 20] \times [0, 20]$. The number of servers ranges from 1 to 29 with increment of 4, and from 50 to 200 with increment of 30. The number of users ranges from 100 from 1000 with increment of 150. For each number of servers and users, four data are generated. Delay is simply defined by the Euclidean distance, and thus is metric in $S \cup U$. We set the time limit to 120 s for all instances.

Table 2 shows the number of instances that were solved within the time limit of 120 s. Here, "solved" means that the algorithm successfully terminates. The upper table is the summary as the number of users varies. For each entry, the total number of generated instances is 56. As it shows, algo2 and algo1-red solve all the instances, while CPLEX cannot solve some instances with 100 users. The lower table is the summary as the number of servers varies. Again, algo2 and algo1-red solve all the instances, which is also a consequence of the upper table, but CPLEX cannot solve any instances with 110 or more servers. We also see that algo1 cannot solve some instances, and thus we conclude that the heuristic interleaving works pretty well.

Table 2. The number of instances that were solved within the time limit of 120 s.

$|U|$	100	250	400	550	700	850	1000
CPLEX	38	32	31	27	24	23	22
algo1	56	56	56	55	52	47	43
algo2	56	56	56	56	56	56	56
algo1-red	56	56	56	56	56	56	56

$|S|$	1	5	9	13	17	21	25	29	50	80	110	140	170	200
CPLEX	28	28	28	28	28	25	15	11	4	2	0	0	0	0
algo1	28	28	28	28	28	28	28	28	28	28	27	23	20	15
algo2	28	28	28	28	28	28	28	28	28	28	28	28	28	28
algo1-red	28	28	28	28	28	28	28	28	28	28	28	28	28	28

---

[1] The reader may wonder why the authors did not make a semi-log plot, which could show the trend better. However, this was impossible since some instances were solved in "0 ms," and taking the logarithm produced $-\infty$.

Figure 3 compares the execution times. The left figure corresponds to the case with 50 servers, the right figure corresponds to the case with 200 servers, and the horizontal axis refers to the number of users. As they show, algo2 and algo1-red scale very well, and even when the number of users is 1000, they run within ten milliseconds.

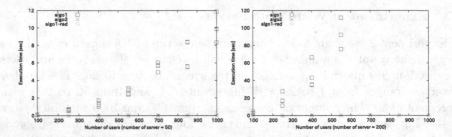

Fig. 3. (Left) Comparison of execution times in Experiment 2, for the case with 50 servers. (Right) The same plot with 200 servers.

Figure 4 compares the objective values. The plot is restricted to the instances that CPLEX was able to solve since we need to know the optimal value for plotting the approximation ratio. The left figure corresponds to the case with 25 servers, where the horizontal axis refers to the number of users. The right figure corresponds to the case with 1000 users, where the horizontal axis refers to the number of servers. As they show, algo2 tends to output worse solutions than algo1 and algo1-red, but the approximation ratio never exceeds 1.15. Our inspection shows that the approximation ratios stay below 1.19 for all the instances that were solved by CPLEX.

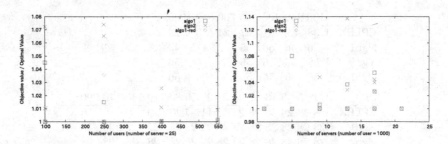

Fig. 4. (Left) Comparison of objective values in Experiment 2, for the case with 25 servers with different numbers of users. (Right) The same plot with 1000 users with different numbers of servers. Each point corresponds to a single instance that was solved by CPLEX.

Acknowledgments. We thank Eiji Oki for bringing the problem into our attention.

References

1. Alumur, S., Kara, B.Y.: Network hub location problems: the state of the art. Eur. J. Oper. Res. **190**(1), 1–21 (2008)
2. Ba, S., Kawabata, A., Chatterjee, B.C., Oki, E.: Computational time complexity of allocation problem for distributed servers in real-time applications. In: Proceedings of 18th Asia-Pacific Network Operations and Management Symposium, pp. 1–4 (2016)
3. Byrka, J., Pensyl, T., Rybicki, D., Srinivasan, A., Trinh, K.: An improved approximation for k-median, and positive correlation in budgeted optimization. In: Proceedings of the Twenty-Sixth Annual ACM-SIAM Symposium on Discrete Algorithms, pp. 737–756 (2015)
4. Campbell, J.F.: Integer programming formulations of discrete hub location problems. Eur. J. Oper. Res. **72**(2), 387–405 (1994)
5. Chen, L.-H., Cheng, D.-W., Hsieh, S.-Y., Hung, L.-J., Lee, C.-W., Wu, B.Y.: Approximation algorithms for single allocation k-hub center problem. In: Proceedings of the 33rd Workshop on Combinatorial Mathematics and Computation Theory (CMCT 2016), pp. 13–18 (2016)
6. Farahani, R.Z., Hekmatfar, M., Arabani, A.B., Nikbakhsh, E.: Hub location problems: a review of models, classification, solution techniques, and applications. Comput. Ind. Eng. **64**(4), 1096–1109 (2013)
7. Gonzalez, T.F.: Clustering to minimize the maximum intercluster distance. Theor. Comput. Sci. **38**, 293–306 (1985)
8. Hochbaum, D.S., Shmoys, D.B.: A best possible heuristic for the k-center problem. Math. Oper. Res. **10**(2), 180–184 (1985)
9. Impagliazzo, R., Paturi, R.: On the complexity of k-SAT. J. Comput. Syst. Sci. **62**(2), 367–375 (2001)
10. Impagliazzo, R., Paturi, R., Zane, F.: Which problems have strongly exponential complexity? J. Comput. Syst. Sci. **63**(4), 512–530 (2001)
11. Kawabata, A., Chatterjee, B.C., Oki, E.: Distributed processing communication scheme for real-time applications considering admissible delay. In: Proceedings of 2016 IEEE International Workshop Technical Committee on Communications Quality and Reliability, pp. 1–6 (2016)
12. O'Kelly, M.E., Miller, H.J.: Solution strategies for the single facility minimax hub location problem. Pap. Reg. Sci. **70**(4), 367–380 (1991)

Computing with Risk and Uncertainty

Risk Aware Stochastic Placement of Cloud Services: The Case of Two Data Centers

Galia Shabtai[1]([✉]), Danny Raz[2], and Yuval Shavitt[1]

[1] School of Electrical Engineering, Tel Aviv University,
Ramat Aviv, 69978 Tel Aviv, Israel
galiashabtai@gmail.com, shavitt@eng.tau.ac.il
[2] Faculty of Computer Science, The Technion, 32000 Haifa, Israel
danny@cs.technion.ac.il

Abstract. Allocating the right amount of resources to each service in any of the data centers in a cloud environment is a very difficult task. This task becomes much harder due to the dynamic nature of the workload and the fact that while long term statistics about the demand may be known, it is impossible to predict the exact demand in each point in time. As a result, service providers either over allocate resources and hurt the service cost efficiency, or run into situation where the allocated local resources are insufficient to support the current demand. In these cases, the service providers deploy overflow mechanisms such as redirecting traffic to a remote data center or temporarily leasing additional resources (at a higher price) from the cloud infrastructure owner. The additional cost is in many cases proportional to the amount of overflow demand.

In this paper we propose a stochastic based placement algorithm to find a solution that minimizes the expected total cost of ownership in case of two data centers. Stochastic combinatorial optimization was studied in several different scenarios. In this paper we extend and generalize two seemingly different lines of work and arrive at a general approximation algorithm for stochastic service placement that works well for a very large family of overflow cost functions. In addition to the theoretical study and the rigorous correctness proof, we also show using simulation based on real data that the approximation algorithm performs very well on realistic service workloads.

1 Introduction

The recent rapid development of cloud technology gives rise to many-and-diverse services being deployed in datacenters across the world. The placement of services to the available datacenters in the cloud has a critical impact on the ability to provide a ubiquitous cost-effective high quality service. There are many challenges associated with optimal service placement due to the large scale of the problem, the need to obtain state information, and the geographical spreading of the datacenters and users.

This paper was supported in part by the Neptune Consortium, Israel and by the Israeli Ministry of Science, Technology and Space.

© Springer International Publishing AG, part of Springer Nature 2018
D. Alistarh et al. (Eds.): ALGOCLOUD 2017, LNCS 10739, pp. 59–88, 2018.
https://doi.org/10.1007/978-3-319-74875-7_5

One intriguing problem is the fact that resource requirement of services changes over time and is not fully known at the time of placement. Moreover, while the average demand may follow a clear daily pattern, the actual demand of a service at a specific time may vary considerably according to the stochastic nature of the demand. One way of addressing this important problem is over-provisioning, that is, allocating resources according to the peak demand. Clearly, this is not a cost effective approach and much of the resources are unused most of the time. A more economical approach, relying on the stochastic nature of the demand, is to allocate just the right amount of resources and potentially use additional mechanisms (such as diverting the service request to a remote location or dynamically buying additional resources) in case of overflow situations where demand exceeds the capacity. Clearly, the cost during such (overflow) events is higher than the normal cost. Moreover, in many cases it is proportional to the amount of unavailable resources. Obviously, the quantitative way of modeling the cost of an overflow situation, considerably depends on the actions taken (or not taken) in such cases. For example one may want to minimize the probability of an overflow event, while another to minimize the expected overflow of the demand. The challenge is therefore to find an optimal placement for a given cost function.

The problem we are dealing with falls into the framework of *stochastic combinatorial optimization*, which has a large body of work in the stochastic optimization literature. Kleinberg et al. [1] were the first to suggest the stochastic load balancing, stochastic bin packing (SBP) and stochastic knapsack problems in the context of bursty connections. They mostly considered Bernoulli-type distributions. Goel and Indyk [2] further studied these problems with Poisson and Exponential distributions. Later, Wang et al. [3] as well as Breitgand and Epstein [4], who considered consolidation of virtual machines in data centers, studied the stochastic bin packing with Normal distributions.

In another line of work, Nikolova et al. [5] considered the stochastic shortest path problem, where one tries to find a path between two points on a graph maximizing the probability of reaching the destination within a given timeframe. Nikolova [6] generalized this problem to other risk-averse stochastic problems with a quasi-concave minimization function. The techniques used in this line of work are very different from those used in [1–4].

We concentrate on three stochastic optimization problems. The first problem is the SP-MED problem (stochastic placement with minimum expected deviation) where our goal is to partition the set of services into two data centers mimizing the overall expected deviation. The other two problems are SP-MWOP (stochastic placement with minimum worst overflow probability) and SP-MOP (stochastic placement with minimum overflow probability). The *exact* version of the problems is NP-hard so our goal is to find an *approximate* solution. The cost functions in these problems are *not* quasi-concave so these problems do not fall into the framework developed by [5,6].

The case of two data centers in the cloud, is quite challenging and in current work in progress we show it is key for solving the general $k > 2$ data centers case [7]. Following Breitgand and Epstein [4] we look at the variance to mean ratio. We think of the amount of variance per one unit of expectation as a risk associated with each service and prove that the optimal solution for two data centers is obtained by putting all the low risk services in one data center, and all the high risk services in the other. Intuitively, this happens because it is beneficial to give the high risk services as much spare capacity as possible, and we achieve that by grouping all the low risk services together and giving them less spare capacity.

The correctness proof partially falls into the framework developed by Nikolova et al. [5]. As in [5], we start with the observation that when the input describes a stochastic behavior of independent Normal distributions, the optimization problem can be reduced to a problem in two dimensions only, where every possible partition corresponds to a feasible point in the plane, and the cost function is a function of two variables only (see Sect. 2). This is because a Normal distribution is captured by its mean and variance, and both the mean and the variance are *additive* when applied on a sum of independent Normal distributions. Thus, we can decouple the optimization problem into two separate and almost orthogonal problems: the first is understanding the feasible set of discrete solutions, and the second is the behavior of the objective cost function as a *continuous* function over the two-dimensional domain.

Nikolova et al. [5,6] study a problem where the feasible set of solutions is a two-dimensional polygon and the cost function is quasi-concave, which implies that the optimum lies on a vertex of the polygon. A major challenge [5,6] face is that in their case determining all the vertices of the polygon of feasible sets, or even just approximating the vertices, is NP-hard. [5,6] show that for minimization of quasi-concave cost functions one can concentrate on a specific part of the boundary which can be determined in polynomial time.

In our case the underlying polygon is the convex hull of all possible partitions. We show that this polygon has a very nice structure and its boundary can be determined in close to linear time (see Sect. 2). Thus, the main difficulty dealt with in [5,6] does not exist in our case. As a result, we can deal with a much wider class of cost functions, and in particular with cost functions that have their optimum on an *arbitrary* point on the boundary. For example, all the three cost functions we consider are *not* quasi-concave and therefore do not fall into the [5,6] framework. In Sect. 2 we define a wide class of functions that falls into our framework and includes many natural optimization functions. We thus extend and benefit from the two separate lines of works described above. We believe the new framework developed in this paper is applicable to many natural resource allocation problems.

We remark that [1–4] study some nicely behaved input distributions like Normal, Poisson, Exponential, Bernoulli and others. Our work is not limited to stochastic Normal distributions only. We use a quantitative version of the central limit theorem to prove that our algorithm works for arbitrary demand distributions as long as the number of services is large and the first three moments of

the services satisfy a mild condition. This is because in such a situation the sum
of many independent random variables converges to a Normal distribution.

We complement the theoretical analysis with a simulation based performance
study in realistic scenarios. We implemented our algorithm and evaluated the
performance over real data obtained from a mid-size operational data center
and emulated workloads. Our results indicate that the new algorithm achieves
a considerable gain when compared with commonly used naive solutions.

2   The Normal Two Bin Case

Following is a general formulation of the problem, with arbitrary number of
data centers (bins). In this paper we solve the two bin case, while the more
general case (of more than two bins) is studied in [7]. The input to the problem
consists of k and n, specifying the number of bins and services, and integers
$\{c_j\}_{j=1}^k$, specifying the bin capacities. We are also given a partial description of
n *independent* random variables $X = (X^{(1)}, \ldots, X^{(n)})$. This partial description
includes the mean $\mu^{(i)}$, variance $V^{(i)} = \mathbb{E}(|X^{(i)} - \mu^{(i)}|^2)$ and $\rho^{(i)} = \mathbb{E}(|X^{(i)} -
\mu^{(i)}|^3)$ of each variable $X^{(i)}$ ($\rho^{(i)}$ are needed only for error estimation). The
output is a partition of $[n]$ to k disjoint sets $S_1, \ldots, S_k \subseteq [n]$, where S_j includes
the indices of the services that are allocated to bin j[1]. Our goal is to find a
partition minimizing the SP-MED[2] cost function, i.e., to find a partition $S =
\{S_1, \ldots, S_k\}$ that obtains the minimal expected deviation. The cost function D_X
is defined as:

$$\mathbf{D_X(S)} = \sum_{j=1}^k \mathbb{E}\mathbf{f_j(X_j)}$$

where X_j is the sum over all services placed in bin j, i.e., $X_j = \sum_{i \in S_j} X^{(i)}$, $f_j(x)$
is the deviation function of bin j, i.e., $f_j(x) = x - c_j$ if $x > c_j$ and 0 otherwise,
and $\mathbb{E}f_j(X_j)$ is the expected deviation of bin j.

An important special case is when each $X^{(i)}$ is *normally distributed* with
mean $\mu^{(i)}$ and variance $V^{(i)}$, and then we denote the cost by $D_N(S)$. In the
normal case we have an explicit formula for $D_N(S)$, namely,

$$D_N(S) = \sum_{j=1}^k Dev_{S_j},$$

---

[1] In this case, the goal is to find an *integral* solution, in which services cannot be split
between bins. Later on, we also consider *fractional* solutions, which allow splitting
a service between several bins.

[2] While we present the results only for SP-MED, we try to keep the discussion in this
section as general as possible, so that it is clear what properties are required from a
cost function to fall into our framework.

where,

$$Dev_{S_j} = \frac{1}{\sigma_j\sqrt{2\pi}} \int_{c_j}^{\infty} (x - c_j)e^{-\frac{(x-\mu_j)^2}{2\sigma_j^2}} dx \tag{1}$$

$$= \sigma_j[\phi(\Delta_j) - \Delta_j(1 - \Phi(\Delta_j))], \tag{2}$$

ϕ is the probability density function of the standard normal distribution and Φ is its cumulative distribution function. Also, $\mu_j = \sum_{i \in S_j} \mu^{(i)}$, $\sigma_j = \sqrt{V_j} = \sqrt{\sum_{i \in S_j} V^{(i)}}$ and $\Delta_j = \frac{c_j - \mu_j}{\sigma_j}$.

In the two bin case the input is $c_1, c_2, \{\mu^{(i)}, V^{(i)}\}_{i=1}^{n}$ as before. If we take a partition $S = (S_1, S_2)$ then at the j'th bin (for $j = 1, 2$) we get the distribution $\sum_{i \in S_j} X^{(i)}$ which is normally distributed with mean $\mu_j = \sum_{i \in S_j} \mu^{(i)}$ and variance $V_j = \sum_{i \in S_j} V^{(i)}$. The cost function is a function of $(\mu_1, V_1), (\mu_2, V_2)$. Notice that $\mu_1 + \mu_2 = \mu = \sum_i \mu^{(i)}$ and $V_1 + V_2 = V = \sum_i V^{(i)}$. Therefore, the cost function depends only on (μ_1, V_1).

We define a function $D : [0, 1]^2 \to \mathbf{R}$ where $D(a, b)$ is the cost function D_N under a partition where the demand to the first bin is normally distributed with mean $a\mu$ and variance bV.

For the cost function SP-MED, $D(a, b) = Dev_{S_1} + Dev_{S_2}$, where as in Eq. (2), Dev_1 depends on $\sigma_1 = \sqrt{bV}$ and $\Delta_1 = \frac{c_1 - a\mu}{\sqrt{bV}}$, and Dev_2 depends on $\sigma_2 = \sqrt{(1-b)V}$ and $\Delta_2 = \frac{c_2 - (1-a)\mu}{\sqrt{(1-b)V}}$. We shall prove (in Appendix A) that the function $D(a, b)$ has the following properties:

1. Symmetry: $D(a, b) = D(1 - a - \frac{c_2 - c_1}{\mu}, 1 - b)$. When $c_1 = c_2$ this simply translates to $D(a, b) = D(1 - a, 1 - b)$ which means that there is no difference between allocating the set S_1 to the first bin or to the second one.
2. Uni-modality in a: For every fixed $b \in [0, 1]$, $D(a, b)$ has a unique minimum on $a \in [0, 1]$, at some point $a = m(b)$, i.e. D is decreasing at $a < m(b)$ and increasing at $a > m(b)$. We call the points on the curve $\{(m(b), b)\}$ the valley.
3. Central saddle point: D has a unique maximum over the valley at the point $(m(\frac{1}{2}), \frac{1}{2})$. In fact since D is symmetric this point has to be $(\frac{1}{2} - \frac{c_2 - c_1}{2\mu}, \frac{1}{2})$. This means that $D(m(b), b)$ is decreasing for $b \leq \frac{1}{2}$ and increasing for $b \geq \frac{1}{2}$.

We will show that these three properties are true for a very large family of cost functions, and in particular for three cost functions that are often used in practice (see Sect. 3). The remarkable fact is that there is a single algorithm that provably works well for every D that has the above three properties.

2.1   The Sorting Algorithm

The sorting algorithm

- Sort the bins by their capacity such that $c_1 \leq c_2$.
- Sort the services by their variance to mean ratio (VMR), i.e., $\frac{V^{(1)}}{\mu^{(1)}} \leq \frac{V^{(2)}}{\mu^{(2)}} \leq \cdots \leq \frac{V^{(n)}}{\mu^{(n)}}$.
- Define $P^{(i)} = (\frac{\mu^{(i)}}{\mu}, \frac{V^{(i)}}{V})$, $P^{[i]} = P^{(1)} + \ldots + P^{(i)}$ and in addition define $P^{[0]} = (0,0)$. Notice that $P^{[n]} = (1,1)$.
- Calculate $Dev(P^{[i]})$ for each $0 \leq i \leq n$ and find the index i^* such that the point $P^{[i^*]}$ achieves the minimal cost among all points $P^{[i]}$.
- Output $(S_1 = \{1, \ldots, i^*\}, S_2 = [n] \setminus S_1)$.

We assume that no input service is too *dominant*. Recall that $P^{(1)} + P^{(2)} + \ldots + P^{(n)} = (1,1)$. Thus, $\sum_i |P^{(i)}| \geq |(1,1)| = \sqrt{2}$ (by the triangle inequality) and $\sum_i |P^{(i)}| \leq 2$ (because the length of the longest increasing path from $(0,0)$ to $(1,1)$, is obtained by the path going from $(0,0)$ to $(1,0)$ and then to $(1,1)$). Hence, the average length of an input point $P^{(i)}$ is somewhere between $\frac{\sqrt{2}}{n}$ and $\frac{2}{n}$. The above assumption (that no input service is too *dominant*) states that no element takes more than L times its "fair" share, i.e., that for some $L \geq 0$, $|P^{(i)}| \leq \frac{L}{n}$ for every i. We also let α denote the (normalized) total spare capacity, i.e. $\alpha = \frac{c-\mu}{\mu}$. Our working assumption is that α is some positive constant. With that we prove:

Theorem 1. *The difference between the cost found by the sorting algorithm and the optimal integral (or fractional) cost is at most $O(\frac{L}{\alpha n})$.*

In the next subsection we give an informal proof of the theorem. A formal proof appears in Appendix E.1.

2.2   The Correctness Proof

We begin with a geometric interpretation of the space of feasible (integral or fractional) solutions. If we split the services according to the partition $S = (S_1 = I, S_2 = [n] \setminus I)$, then the first bin is normally distributed with mean $\mu \sum_{i \in I} a^{(i)}$ and variance $V \sum_{i \in I} b^{(i)}$, where $a^{(i)} = \frac{\mu^{(i)}}{\mu}$ and $b^{(i)} = \frac{V^{(i)}}{V}$. Thus, our cost is $D(P_I)$ where $P_I = \sum_{i \in I} P^{(i)}$ and $P^{(i)} = (a^{(i)}, b^{(i)})$. We call each such point *an integral point*. Sorting the services by their VMR, is equivalent to sorting the vectors $P^{(i)}$ by the angle they make with the a axis.

Definition 1. *(The sorted paths) Sort the services by their VMR in increasing order and calculate the $P^{(1)}, P^{(2)}, \ldots, P^{(n)}$ vectors. For $i = 1, \ldots, n$ define*

$$P^{[i]}_{bottom} = P^{(1)} + P^{(2)} + \ldots + P^{(i)} \text{ and,}$$
$$P^{[i]}_{up} = P^{(n)} + P^{(n-1)} + \ldots + P^{(n-i+1)},$$

and also define $P^{[0]}_{bottom} = P^{[0]}_{up} = (0,0)$.

The bottom sorted path *is the curve that is formed by connecting* $P_{bottom}^{[i]}$
and $P_{bottom}^{[i+1]}$ *with a line, for* $i = 0, \ldots, n-1$. *The* upper sorted path *is the curve*
that is formed by connecting $P_{up}^{[i]}$ *and* $P_{up}^{[i+1]}$ *with a line, for* $i = 0, \ldots, n-1$.

The integral point $P_{bottom}^{[i]}$ on the bottom sorted path corresponds to allocating the i services with the lowest VMR to the first bin and the rest to the second. Similarly, the integral point $P_{up}^{[i]}$ on the upper sorted path corresponds to allocating the i services with the highest VMR to the first bin and the rest to the second. A crucial, yet simple, observation:

Lemma 1. *All the integral points lie within the polygon confined by the bottom sorted path and the upper sorted path.*

Proof. We introduce some notation. Let $\tau = \tau_1, \ldots, \tau_n$ be a sequence of n elements that is a permutation of $\{1, \ldots, n\}$. We associate with τ the n partial sums $P_\tau^{[1]}, \ldots, P_\tau^{[n]}$ where $P_\tau^{[i]}$ is $\sum_{j=1}^{i} P^{(\tau_j)}$, i.e., $P_\tau^{[i]}$ is the integral point that is the sum of the first i points according to the sequence τ. We also define $P_\tau^{[0]} = (0,0)$ and $P_\tau^{[n]} = (1,1)$. The *curve connecting* τ is the curve that is formed by connecting $P_\tau^{[i]}$ and $P_\tau^{[i+1]}$ with a line, for $i = 0, \ldots, n-1$.

Assume that in the sequence $\tau = \tau_1, \ldots, \tau_n$ there is some index i such that the VMR of $P^{(\tau_i)}$ is larger than the VMR of $P^{(\tau_{i+1})}$. Consider the sequence τ' that is the same as τ except for switching the order of τ_i and τ_{i+1}. I.e., $\tau' = \tau_1, \ldots, \tau_{i-1}, \tau_{i+1}, \tau_i, \tau_{i+2}, \ldots, \tau_n$. We claim that the curve connecting τ' lies beneath the curve connecting τ. To see that notice that both curves are the same up to the point $P_\tau^{[i-1]}$. There, the two paths split. τ adds $P^{(\tau_i)}$ and then $P^{(\tau_{i+1})}$ while τ' first adds $P^{(\tau_{i+1})}$ and then $P^{(\tau_i)}$. Then the two curves coincide and overlap all the way to $(1,1)$. In the section where the two paths differ, the two different paths form a parallelogram with $P^{(\tau_i)}$ and $P^{(\tau_{i+1})}$ as two neighboring edges of the parallelogram. As the angle $P^{(\tau_{i+1})}$ has with the a axis is smaller than the angle $P^{(\tau_i)}$ has with the a axis, the curve connecting τ' lies beneath that of τ.

To finish the argument, let P_I be an arbitrary integral point for some $I \subseteq [n]$. Look at the sequence τ that starts with the elements of I followed by the elements of $[n] \setminus I$ in an arbitrary order. Notice that P_I lies on the curve connecting τ. Now run a bubble sort on τ, each time ordering a pair of elements by their VMR. Notice that the process terminates with the sequence that sorts the elements by their VMR and the curve connecting the final sequence is the bottom sorted path. Thus, we see that the bottom sorted path lies beneath the curve connecting τ, and in particular P_I lies above the bottom sorted path. A similar argument shows P_I lies underneath the upper sorted path.

We can say more. A fractional partition is one that allows splitting a service between several bins. Geometrically, the set of fractional points is a convex set. Clearly, it contains all the points on both the bottom sorted line and the upper sorted line, and because it defines a convex set, also all points in their convex hull. In fact,

Lemma 2. *The set of fractional points coincides with the polygon confined by the bottom sorted path and the upper sorted path.*

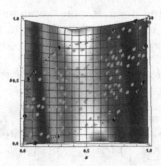

Fig. 1. The figure depicts $D(a, b)$ when $\mu = 160$, $V = 6400$, $c_1 = c_2 = 100$ and the cost function is $SP - MED$. The orange points are the 2^{10} integral partition points. The dotted lines are the bottom and upper sorted paths. Notice that all the integral partition points are confined by the bottom and upper sorted paths. (Color figure online)

Figure 1 demonstrates such a polygon. Having this geometric picture we prove:

Theorem 2. *The optimal fractional point lies on the bottom sorted path. The optimal fractional solution splits at most one service between two bins.*

Proof. Consider an arbitrary fractional point (a_0, b_0) lying strictly inside the polygon confined by the upper and bottom sorted paths. If $b_0 \leq \frac{1}{2}$, then by keeping $b = b_0$ constant and changing a till it reaches the valley we strictly decrease cost (because D is strictly monotone in this range). Now, when changing a we either hit the bottom sorted path or the valley. If we hit the bottom sorted path, we found a point on the bottom sorted path with less cost and we are done. If we hit the valley, we can go down the valley until we hit the bottom sorted path and again we are done (as D is strictly monotone on the valley).

We now consider the case $b_0 \geq \frac{1}{2}$. Notice that if (α, β) is an integral point on the upper sorted path induced by the partition $I \subseteq [n]$, then the integral point induced by $[n] \setminus I$ is $(1 - \alpha, 1 - \beta)$ and it lies on the bottom sorted path. The same holds in the reverse direction. In particular the mapping $\varphi : [0, 1]^2 \rightarrow [0, 1]^2$ defined by $\varphi(a, b) = (1 - a, 1 - b)$ maps fractional points to fractional points and integral points to integral points, the upper sorted path to the bottom sorted path and vice versa. An example can be seen in Fig. 1. Note that the points (a, b) and $\varphi(a, b)$ might have different costs when $c_1 \neq c_2$, and the symmetry condition only guarantees $D(a, b) = D(1 - a - \frac{c_2 - c_1}{\mu}, 1 - b)$. Then,

- The point $(1 - a_0, 1 - b_0) = \varphi(a_0, b_0)$ is fractional (since (a_0, b_0) is fractional and φ maps fractional points to fractional points), and,
- By the reflection symmetry we know that $D(a_0, b_0) = D(1 - a_0 - \zeta, 1 - b_0)$ where $\zeta = \frac{c_2 - c_1}{\mu} \geq 0$.

Now, $(1 - a_0 - \zeta, 1 - b_0)$ has b coordinate that is at most $\frac{1}{2}$. Also $(1 - a_0 - \zeta, 1 - b_0)$ lies to the *left* of the fractional point $(1 - a_0, 1 - b_0)$ (since $\zeta > 0$) and therefore it lies above the bottom sorted path. We therefore see that the point (a_0, b_0) has a corresponding fractional point with the same cost and with b coordinate at most $\frac{1}{2}$. Applying the argument that appears in the first paragraph of the proof we conclude that there exists some point on the bottom sorted path with less cost, and conclude the proof.

In the introduction we said that the optimal solution allocates low risk services to one bin and the rest to the other. However, when $c_1 \neq c_2$ it is not clear whether to allocate the smaller risk services to the lower capacity bin or the higher capacity bin. Equivalently, offhand, it is not clear whether the optimal solution lies on the bottom sorted path or the upper sorted path, and it might even depend on the input. Theorem 2 proves that when $c_1 \leq c_2$ the optimal solution lies on the bottom sorted path, meaning that it is always better to allocate the low risk services to the smaller capacity bin and the high risk services to the higher capacity bin.

Figure 2 depicts $D(a, b)$ for SP-MED. From looking at the left figure one gets the impression that the saddle point $(\frac{1}{2}, \frac{1}{2})$ is the optimal solution. However, a close-up around this saddle point reveals that there is a much better solution that can be obtained by going down the "valley", and in fact the point $(\frac{1}{2}, \frac{1}{2})$ is the highest point on that valley.

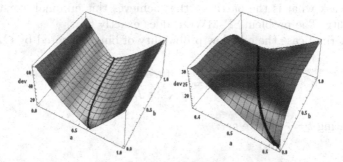

Fig. 2. We again consider $D(a, b)$ when $\mu = 160$, $V = 6400$, $c_1 = c_2 = 100$ for $SP - MED$. Looking at the left figure one gets the impression the saddle point $(\frac{1}{2}, \frac{1}{2})$ is optimal. However, the right figure is a zoom in around the saddle point $(\frac{1}{2}, \frac{1}{2})$ and clearly shows there are much better solutions down the valley (marked by a black line).

What is left now is estimating the errors made by the algorithm. The sorting algorithm finds an *integral* point on the bottom sorted path, and its cost should be compared with the value of the best *fractional* point on the bottom sorted path. By our assumption on L the integral points form a dense net on the bottom sorted path. Using that and standard tools like the mean value theorem for multi-variate functions we get our error estimate. The proofs are technical and we omit them. The full details can be found in Appendix E.

3   Other Cost Functions

We present two more cost functions that fall into our framework.

- **SP-MWOP** (*Stochastic Placement with Min Worst Overflow Probability*): In SP-MWOP the cost is the minimal probability p, such that for every bin the probability that the bin overflows is at most p. Namely, if OF_j is the event that bin j overflows, then the cost of a placement is $\max_{j=1}^{k} \Pr[\mathbf{OF_j}]$. The SP-MWOP problem gets as input integers k and n, specifying the number of bins and services, integers $c_1, \ldots, c_k$, specifying the bin capacities and values $\left\{ (\mu^{(i)}, V^{(i)}) \right\}_{i=1}^{n}$, specifying that the demand distribution $X^{(i)}$ of service i is normal with mean $\mu^{(i)}$ and variance $V^{(i)}$. A solution to the problem is a partition of $[n]$ to k disjoint sets $S_1, \ldots, S_k \subseteq [n]$ that minimizes the worst overflow probability.

  The SP-MWOP problem is a natural variant of SBP. For a given partition let OFP_j (for $j = 1, \ldots, k$) denote the overflow probability of bin j. Let $WOFP$ denote the *worst* overflow probability, i.e., $WOFP = \max_{j=1}^{k} \{OFP_j\}$. In the SBP problem we are given n normal distributions and wish to pack them into few bins such that the $OFP \le p$ for some given parameter p. Suppose we solve the SBP problem for a given p and know that k bins suffice. We now ask ourselves what is the minimal $WOFP$ achieved with the k bins (this probability is clearly at most p but can also be significantly smaller). We also ask what is the partition that achieves this minimal worst overflow probability. The problem SP-MWOP does exactly that.

  In the normal case the overflow probability of bin j, denoted by OFP_j, is:

$$OFP_j(\mu_j, V_j) = \frac{1}{\sigma_j \sqrt{2\pi}} \int_{c_j}^{\infty} e^{-\frac{(x - \mu_j)^2}{2\sigma_j^2}} \, dx$$

  Substituting $t = \frac{x - \mu_j}{\sigma_j}$ we get:

$$OFP_j(\mu_j, V_j) = \frac{1}{\sqrt{2\pi}} \int_{\frac{c_j - \mu_j}{\sigma_j}}^{\infty} e^{-\frac{t^2}{2}} \, dt$$

$$= 1 - \Phi(\frac{c_j - \mu_j}{\sigma_j}) \;=\; 1 - \Phi(\Delta_j).$$

  Thus,

$$WOFP = \max_{j=1}^{k} \{1 - \Phi(\Delta_j)\}.$$

With two bins $WOFP$ is a function from $[0,1]^2$ to $\mathbf{R}$ and,

$$WOFP(a,b) = \max\{1 - \Phi(\Delta_1), 1 - \Phi(\Delta_2)\}$$

where the first bin has mean $a\mu$ and variance bV, the second bin has mean $(1-a)\mu$ and variance $(1-b)V$. σ_j, Δ_j were previously defined.

– **SP-MOP** (*Stochastic Placement with Minimum Overflow Probability*): In SP-MOP the cost is the probability that *any* bin overflows, i.e. $\Pr[\bigcup_{j=1}^{k} \mathbf{OF_j}]$. The SP-MOP problem gets as input integers k and n, specifying the number of bins and services, integers $c_1, \ldots, c_k$, specifying the bin capacities and values $\{(\mu^{(i)}, V^{(i)})\}_{i=1}^{n}$, specifying that the demand distribution $X^{(i)}$ of service i is normal with mean $\mu^{(i)}$ and variance $V^{(i)}$. A solution to the problem is a partition of $[n]$ to k disjoint sets $S_1, \ldots, S_k \subseteq [n]$ that minimizes the overflow probability.

The total overflow probability is $OFP = 1 - \prod_{j=1}^{k}(1 - OFP_j)$ where in the normal case, as computed before, $OFP_j = 1 - \Phi(\Delta_j)$. With two bins OFP is a function from $[0,1]^2$ to $\mathbf{R}$ and $OFP(a,b) = 1 - \Phi(\Delta_1)\Phi(\Delta_2)$ where the first bin has mean $a\mu$ and variance bV, the second bin has mean $(1-a)\mu$ and variance $(1-b)V$ and σ_j, Δ_j were previously defined.

We prove in Appendices B and C that both SP-MWOP and SP-MOP fall into our framework:

Theorem 3. *OFP and WOFP respect the symmetry, uni-modality and the central saddle point property.*

Hence, by Theorem 2 we know that the optimal fractional solution is obtained on the bottom sorted path. In fact, for SP-MWOP we can say a bit more, and in Appendix B we prove:

Theorem 4. *The optimal fractional solution for SP-MWOP is the unique point that is the intersection of the valley and the bottom sorted path and in this point $\Delta_1 = \Delta_2$.*

4 Non-normal Distributions

4.1 The Berry-Esseen Theorem

The Kolmogorov distance between two cumulative distribution functions F and G is given by $\|F - G\|_\infty = \sup_{t \in \mathbb{R}} |F(t) - G(t)|$. The Central Limit Theorem states that the sum of independent arbitrary random variables converges (when the number of random variables tends to infinity) to the Normal distribution. The convergence is in the Kolmogorov distance. The Berry-Esseen theorem is a quantitative version of the Central Limit Theorem, giving a quantitative bound on the rate of convergence.

Theorem 5 *(Berry-Esseen). Let $X^{(1)}, \ldots, X^{(n)}$ be independent random variables with*

$$\mu^{(i)} = \mathbb{E}(X^{(i)}),$$
$$V^{(i)} = \mathbb{E}(|X^{(i)} - \mu^{(i)}|^2),$$
$$\rho^{(i)} = \mathbb{E}(|X^{(i)} - \mu^{(i)}|^3).$$

Let F_N denote the cumulative distribution function of $N(\mu, V)$ for $\mu = \sum \mu^{(i)}$ and $V = \sum V^{(i)}$. Denote $\sigma = \sqrt{V}$. Let F_X denote the cumulative distribution function of $\sum_{i=1}^{n} X^{(i)}$. Then,

$$\|F_X - F_N\|_\infty \leq C_0 \cdot \psi_0,$$

for $\psi_0 = \frac{\sum_{i=1}^{n} \rho^{(i)}}{V^{\frac{3}{2}}}$ and C_0 some constant in the range $[0.4097, 0.56]$ (see [8, 9]).
Furthermore, for any $t \in \mathbb{R}$:

$$|F_X(t) - F_N(t)| \leq C_1 \cdot \psi_0 \cdot \frac{1}{(\frac{t-\mu}{\sigma})^2 + 1},$$

where C_1 is a universal constant (See [10]).

To say that ψ_0 is small is to simultaneously say two things: the random variables $X^{(i)}$ are all reasonable, in the sense that $\rho^{(i)} \leq O((V^{(i)})^{\frac{3}{2}})$, and none is too dominant in terms of variance [11].

Note that if $X^{(1)}, \ldots, X^{(n)}$ are i.i.d., then $\psi_0 = \frac{\rho^{(1)}}{\sqrt{n}(V^{(1)})^{\frac{3}{2}}}$. As $\rho^{(1)}$ and $V^{(1)}$ are independent of n, we can treat $\rho^{(1)}$ and $V^{(1)}$ as constants and the error goes down to 0 asymptotically with n as $O(n^{-\frac{1}{2}})$.

4.2   Approximating General Independent Distributions with the Normal Distribution

Recall that in Sect. 2 we defined the cost function $D_X(S)$ for general independent random variables $X = (X^{(1)}, \ldots, X^{(n)})$, and $D_N(S)$ for normally distributed random variables. We claim:

Proposition 1. *Given n independent random variables $X = (X^{(1)}, \ldots, X^{(n)})$ with mean $\mu^{(i)}$, variance $V^{(i)} = \mathbb{E}(|X^{(i)} - \mu^{(i)}|^2)$ and $\rho^{(i)} = \mathbb{E}(|X^{(i)} - \mu^{(i)}|^3)$ and a partition $S = \{S_1, \ldots, S_k\}$,*

$$|D_X(S) - D_N(S)| \leq C_1 \sum_{j=1}^{k} \sigma_j \psi_0^j(S_j)(\frac{\pi}{2} - \arctan(\Delta_j)),$$

where C_1 is the constant defined in Theorem 5,

$$\psi_0^j(S_j) = \frac{\sum_{i \in S_j} \rho^{(i)}}{V_j^{\frac{3}{2}}},$$

$\mu_j = \sum_{i \in S_j} \mu^{(i)}$, $\sigma_j = \sqrt{V_j} = \sqrt{\sum_{i \in S_j} V^{(i)}}$ and $\Delta_j = \frac{c_j - \mu_j}{\sigma_j}$.

In the proof below we use Fubini's theorem (that in this case can also be derived directly by integration by parts): If X is a non-negative random variable, and F_X is its cumulative distribution function, then

$$E(X) = \int_0^\infty \Pr_X(X \geq t)dt = \int_0^\infty (1 - F_X)(t)dt$$

Proof. Recall that $X_j = \sum_{i \in S_j} X^{(i)}$ and let $N_j = N(\sum_{i \in S_j} \mu^{(i)}, \sum_{i \in S_j} V^{(i)})$. Then:

$$\mathbb{E}f_j(X_j) = \int_0^\infty \Pr_{X_j}(f_j(X_j) \geq t)dt$$

$$= \int_0^\infty \Pr_{X_j}(X_j \geq t + c_j)dt$$

$$= \int_{c_j}^\infty \Pr_{X_j}(X_j \geq t)dt$$

Similarly, $\mathbb{E}f_j(N_j) = \int_{c_j}^\infty \Pr_{N_j}(N_j \geq t)dt$. Therefore,

$$|\mathbb{E}f_j(X_j) - \mathbb{E}f_j(N_j)| = |\int_{c_j}^\infty (F_{X_j} - F_{N_j})(t)dt|$$

$$\leq C_1 \psi_0^j(S_j) \int_{c_j}^\infty \frac{1}{(\frac{t - \mu_j}{\sigma_j})^2 + 1} dt$$

$$= C_1 \psi_0^j(S_j) \int_{\frac{c_j - \mu_j}{\sigma_j}}^\infty \frac{1}{y^2 + 1} \sigma_j dy$$

$$= C_1 \psi_0^j(S_j)\sigma_j(\frac{\pi}{2} - \arctan(\Delta_j)).$$

Finally, $|\sum_j \mathbb{E}f_j(X_j) - \sum_j \mathbb{E}f_j(N_j)| \leq \sum_j |\mathbb{E}f_j(X_j) - \mathbb{E}f_j(N_j)|$ and this completes the proof.

Roughly speaking, Proposition 1 tells us that we do not need to have a complete knowledge of the distribution $X = (X_1, \ldots, X_n)$ but rather that under mild assumptions (namely, that the number of services is large enough for the central limit theorem to hold) it is sufficient to know the first two moments of $X^{(i)}$. Indeed,

Proposition 2. *Let X, D_X, N, D_N be as before. Let S_X (resp. S_N) be the partition in which the optimal solution is achieved under X (resp. N). Suppose that*

$$|D_N(S_X) - D_X(S_X)| \leq \epsilon(S_X)$$
$$|D_N(S_N) - D_X(S_N)| \leq \epsilon(S_N)$$

for some error function $\epsilon(S)$ that may depend on the partition S.[3] Then

$$|D_X(S_X) - D_N(S_N)| \leq \max\{\epsilon(S_X), \epsilon(S_N)\}.$$

---

[3] For example, for SP-MED, $\epsilon(S) = C_1 \sum_{j=1}^k \sigma_j \psi_0^j(S_j)(\frac{\pi}{2} - \arctan(\Delta_j))$.

Proof.

$$D_N(S_N) \leq D_N(S_X)$$
$$\leq D_X(S_X) + \epsilon(S_X), \text{ and,}$$
$$D_N(S_N) \geq D_X(S_N) - \epsilon(S_N)$$
$$\geq D_X(S_X) - \epsilon(S_N).$$

Notice that we need $\epsilon(S)$ to be small only at the two partitions S_N and S_X and we require nothing from all other partitions. What do we expect to see in $\epsilon(S_N)$ and $\epsilon(S_x)$? Luckily, in these two partitions we expect that each bin is allocated many services, and we expect $\psi_0^j(S_j)$ in these partition points to be on the order of about $\frac{1}{\sqrt{n_j}}$, where n_j is the number of services allocated to bin j. Also V_j is the sum of n_j independent bounded random variables, and therefore $\sigma_j = O(\sqrt{n_j})$. Therefore, we expect the error term $\epsilon(S)$ in these two points to be bounded by a constant, independent of n. This is very strong given that in the usual case of interest we expect the cost function (which is the expected overflow) to go to infinity.[4] Thus, using the Berry-Esseen theorem, we get under mild conditions a reduction from the general case, where the independent $X^{(i)}$ are almost arbitrary, to the normal case.

5   Simulation Results

In this section we present our simulation results for the two bin case. We compare the sorting algorithm with two algorithms we call BS (Balanced Spares) and BL (Balanced Load). The BS algorithm goes through the list, item by item, and allocates each item to the bin which has more available space. In this way, the spare capacity is balanced. On the other hand, the BL algorithm goes through the list, item by item, and allocates each item to the bin which is less loaded, i.e., the bin with higher $\frac{\text{available space}}{\text{bin capacity}}$ value. In this way, the bin load is balanced. The BL and BS algorithms are natural benchmarks and also much better than other naive solutions like first-fit and first-fit decreasing.[5] We used several values

---

[4] A similar (simpler and easier to state) result also holds for the other two cost functions we have examined. See Appendix D.

[5] At first, we also wanted to compare our algorithm with variants of the algorithms considered in [3,4] for the SBP problem. In both papers, the authors consider the algorithms First Fit and First Fit Decreasing [12] with item size equal to the effective size, which is the mean value of the item plus an extra value that guarantees an overflow probability is at most some given value p. Their algorithm chooses an existing bin when possible, and otherwise opens a new bin. However, when the number of bins is fixed in advance, taking effective size rather than size does not change much. For a new item (regardless of its size or effective size) we keep choosing the bin that is less occupied, but this time we measure occupancy with respect to effective size rather than size. Thus, if elements come in a random order, the net outcome of this is that the two bins are almost balanced and a new item is placed in each bin with almost equal probability.

for $\frac{c_1}{c}$, i.e. first bin's capacity divided by total capacity (recall that $c = c_1 + c_2$). Note that $0 \leq \frac{c_1}{c} \leq \frac{1}{2}$ (because the sorting algorithm first sorts c_1, c_2 and hence $0 \leq c_1 \leq c_2$).

5.1 Results for Synthetic Normally Distributed Data

We first show simulation results on synthetic *normally distributed* data. We generate the stochastic input $\{(\widetilde{\mu}^{(i)}, \widetilde{\sigma}^{(i)})\}_{i=1}^{n}$ for $n = 500$. Our sample space is a mixture of three populations: all items have the same mean (we fixed it at $\widetilde{\mu}^{(i)} = 500$) but 50% had standard deviation picked uniformly from $[0, 0.4 \cdot \widetilde{\mu}^{(i)}]$, 25% had standard deviation picked uniformly from $[0.4 \cdot \widetilde{\mu}^{(i)}, 0.7 \cdot \widetilde{\mu}^{(i)}]$ and 25% had standard deviation picked uniformly from $[0.7 \cdot \widetilde{\mu}^{(i)}, 0.9 \cdot \widetilde{\mu}^{(i)}]$.

We then randomly generated 800 sample values $x_l^{(i)}$ for each $1 \leq i \leq n$ and $1 \leq l \leq 800$ using the normal distribution $N[\widetilde{\mu}^{(i)}, \widetilde{V}^{(i)}]$ and from this we inferred parameters $\mu^{(i)}, V^{(i)}$, best explaining the sample as a normal distribution. The sorting algorithm, the BS and the BL algorithms got as input $\{(\mu^{(i)}, V^{(i)})\}_{i=1}^{n}$, as well as c_1, c_2 and output their partition.

To check the suggested partitions, we viewed each sample $x_l^{(i)}$ as representing an item instantiation at a different time slot. We then computed the cost function. For example, for SP-MED, the deviation value for bin j at time slot l is: $\max\left\{0, 100 \frac{\sum_{i \in S_j} x_l^{(i)} - c_j}{\sum_{i=1}^{n} \mu^{(i)}}\right\}$, i.e., the deviation is measured as a percent of the total mean value μ. We generated 10 such lists and calculated the average cost for these 10 input lists for each algorithm. We run the experiment for different values of c and for different values of $\frac{c_1}{c}$.

Figure 3 shows the average cost of the three algorithms for SP-MED, SP-MWOP and SP-MOP as a function of $\frac{c}{\mu}$, for $\frac{c_1}{c} \in \{0.1, 0.5\}$. As expected, the average cost decreases as the value $\frac{c}{\mu}$ increases, i.e., as the total spare capacity increases. We also see that the results of the BS and the BL algorithms coincide when $\frac{c_1}{c} = 0.5$, which is obvious. Moreover, the sorting algorithm out-performs the BS and the BL algorithms for both values of $\frac{c_1}{c}$. The advantage of the sorting algorithm is especially evident when $\frac{c_1}{c} = 0.1$. Figure 4 shows the average cost of the BS algorithm divided by the average cost of the sorting algorithm for the three cost functions, as a function of $\frac{c}{\mu}$ for different values of $\frac{c_1}{c}$. When bin capacities are equal (i.e., $\frac{c_1}{c} = 0.5$), the BS algorithm cost is 24.4% (7.0%, 20.4%) higher than the cost of the sorting algorithm for SP-MED (SP-MWOP, SP-MOP, resp.) with 2% spare capacity (i.e., $\frac{c}{\mu} = 1.02$), and 72.2% (57.0%, 75.2%) higher for SP-MED (SP-MWOP, SP-MOP, resp.) with 6% spare capacity (i.e., $\frac{c}{\mu} = 1.06$). The savings get larger when bin capacities are unbalanced (i.e., when $\frac{c_1}{c}$ decreases). For example, when $\frac{c_1}{c} = 0.1$ and the spare capacity is 2%, the BS algorithm cost is 81.3% (47.5%, 70.3%) higher than the cost of the sorting algorithm for SP-MEDpg (SP-MWOP, SP-MOP, resp.). When the spare capacity is 6%, the BS algorithm cost is about 18 (13,11) times the cost of the sorting algorithm for SP-MEDpg (SP-MWOP, SP-MOP, resp.). Figure 5 shows similar and better results (depends

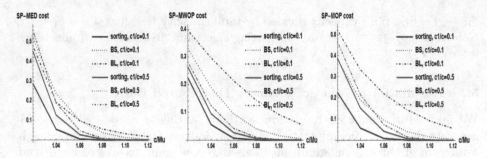

Fig. 3. Average cost of the sorting algorithm and the BS and BL algorithms for SP-MED, SP-MWOP and SP-MOP with two bins for synthetic normally distributed data. The x axis measures $\frac{c}{\mu}$.

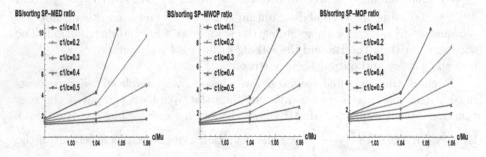

Fig. 4. Average cost of the BS algorithm divided by average cost of the sorting algorithm for SP-MED, SP-MWOP and SP-MOP with two bins for synthetic normally distributed data. The x axis measures $\frac{c}{\mu}$.

on the cost function and the $\frac{c_1}{c}$ value) for the average cost of the BL algorithm divided by the average cost of the sorting algorithm.

5.2   Results for Real Data

In this section, we consider simulation results on real data. We used the real data center trace reported in [4]. It specifies the incoming and outgoing traffic rates for 17 thousand VMs. The distribution of the VM samples is very far from being normal. The standard deviation is higher than the mean value in almost all of the VMs, and it even reaches 10–20 times the mean value.

The number of samples in each VM varies a lot, so we considered only VMs with 800 samples and above (total of 6105 VMs) and took the first 800 receive rate samples from each such VM. For each VM we calculated mean, variance and third moment, for its 800 sample values and from these values we inferred ψ_0, which was very high and impractical (14.11). Therefore, we threw away 6 "problematic" VMs (those with high $\rho^{(i)}/V^{1.5}$ value) and we were left with 6099 VMs and a ψ_0 value of 0.2183.

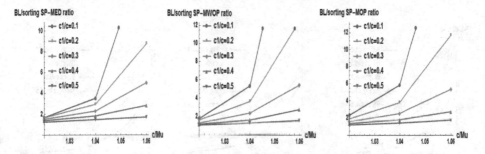

Fig. 5. Average cost of the BL algorithm divided by average cost of the sorting algorithm for SP-MED, SP-MWOP and SP-MOP with two bins for synthetic normally distributed data. The x axis measures $\frac{c}{\mu}$.

Next, since our model assumes independent services, we broke down the dependency between the VMs by taking a random permutation of the 800 VM samples. Since the random permutation only changes the order of the samples, it does not change the statistic values of the mean, variance and third moment nor ψ_0 value. We generated 10 different random permutations for each VM samples and calculated the average cost for these 10 input data sets for each algorithm. We run the experiment for different values of c and for different values of $\frac{c_1}{c}$.

Figure 6 shows the actual average cost of both algorithms for SP-MED, SP-MWOP and SP-MOP as a function of $\frac{c}{\mu}$, for $\frac{c_1}{c} \in \{0.1, 0.5\}$. Again, the average cost decreases as the value $\frac{c}{\mu}$ increases, but not as fast as in the synthetic normal case. As before, we see that the results of the BS and the BL algorithms coincide when $\frac{c_1}{c} = 0.5$, and that for both values of $\frac{c_1}{c}$, the sorting algorithm out-performs the BS and the BL algorithms. The advantage of the sorting algorithm is especially evident when $\frac{c_1}{c} = 0.1$. Figure 7 shows the average cost of the BS algorithm divided by the average cost of the sorting algorithm for the three problems, again as a function of $\frac{c}{\mu}$ for different values of $\frac{c_1}{c}$. We see that the sorting algorithm out-performs the BS algorithm even for this non normally distributed data. When bin capacities are equal (i.e. $\frac{c_1}{c} = 0.5$), the BS algorithm cost is 17.8% (6.8%, 16.2%) higher than the cost of the sorting algorithm for SP-MED (SP-MWOP, SP-MOP, resp.) with 5% spare capacity (i.e., $\frac{c}{\mu} = 1.05$), and 65% (88.8%, 71.4%) higher for SP-MED (SP-MWOP, SP-MOP, resp.) with 25% spare capacity (i.e., $\frac{c}{\mu} = 1.25$). The savings get larger when bin capacities are unbalanced (i.e., when $\frac{c_1}{c}$ decreases). For example, when $\frac{c_1}{c} = 0.1$ and the spare capacity is 5%, the BS algorithm cost is 54.6% (35.5%, 52.8%) higher than the cost of the sorting algorithm for SP-MED (SP-MWOP, SP-MOP, resp.). When the spare capacity is 25%, the BS algorithm cost is about 27 (18,18) times the cost of the sorting algorithm for SP-MED (SP-MWOP, SP-MOP, resp.). Figure 8 shows similar and better results (depends on the cost function and the $\frac{c_1}{c}$ value) for the average cost of the BL algorithm divided by the average cost of the sorting algorithm.

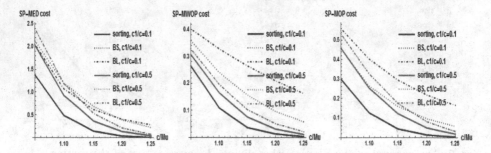

Fig. 6. Average cost of the sorting algorithm and the BS and BL algorithms for SP-MED, SP-MWOP and SP-MOP with two bins for real independent data. The x axis measures $\frac{c}{\mu}$.

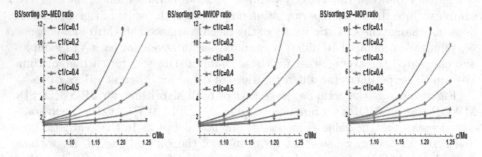

Fig. 7. Average cost of the BS algorithm divided by average cost of the sorting algorithm for SP-MED, SP-MWOP and SP-MOP with two bins for real independent data. The x axis measures $\frac{c}{\mu}$.

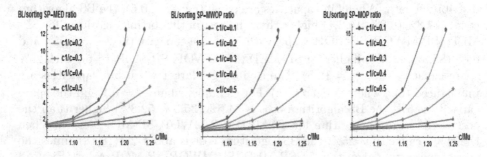

Fig. 8. Average cost of the BL algorithm divided by average cost of the sorting algorithm for SP-MED, SP-MWOP and SP-MOP with two bins for real independent data. The x axis measures $\frac{c}{\mu}$.

6   Conclusions

We present a novel analytical scheme for stochastic placement algorithms, using the stochastic behavior of the demand. We develop efficient, almost optimal algorithms that work for a family of target cost functions. In particular, we solve SP-MED (that minimizes the expected deviation), SP-MOP (that minimizes the probability of overflow) and SP-MWOP (that guarantees that for every bin the probability it overflows is small). We believe the framework is applicable for many other natural cost functions.

Another contribution of this work is its robustness with respect to the input. Much of previous research in the area assumes the services have a particular well-behaved demand distribution (like Bernoulli, [1,2], Exponential [2], Normal [3,4], Poisson [2], etc., to mention a few of the distributions that were considered so far). The results in this paper hold for any large enough collection of independent services of whatever distribution. Furthermore, the amount of robustness can be quantified using the Berry-Esseen theorem, and given stochastic demand one can infer in advance the utility of the methods introduced in the paper.

For every target function and input distribution that fall into the framework, our algorithm examines a linear number of potential solutions and its error decreases fast with the number of services in the input. Our simulation results (see Sect. 5) have obtained a considerable gain over real data from a mid-size operational data center, compared with commonly used naive solutions.

Acknowledgments. We want to thank Liran Rotem for helping us with the proof of Proposition 1 and useful discussion of the Berry-Esseen Theorem and Fubini's Theorem. We also want to thank Boaz Klartag, Ryan O'Donnell and Terry Tao for answering our questions in email, and Ryan for referring us to relevant references.

A   Proving SP-MED Falls into Our Framework

By definition the expected deviation of a single bin is $Dev_{S_j} = \frac{1}{\sigma_j \sqrt{2\pi}} \int_{c_j}^{\infty} (x -$

$c_j) e^{-\frac{(x-\mu_j)^2}{2\sigma_j^2}} dx$. Doing the variable change $t = \frac{x-\mu_j}{\sigma_j}$ and then the variable change $y = \frac{-t^2}{2}$ we get:

$$Dev_{S_j} = (\mu_j - c_j)[1 - \Phi(\frac{c_j - \mu_j}{\sigma_j})] - \frac{\sigma_j}{\sqrt{2\pi}} \int_{-\frac{1}{2}(\frac{c_j-\mu_j}{\sigma_j})^2}^{-\infty} e^y dy$$

$$= -\sigma_j \Delta_j[1 - \Phi(\Delta_j)] + \frac{\sigma_j}{\sqrt{2\pi}} e^{-\frac{1}{2}\Delta_j^2}$$

$$= \sigma_j[\phi(\Delta_j) - \Delta_j(1 - \Phi(\Delta_j))].$$

where ϕ is the probability density function (pdf) of the standard normal distribution and Φ is its cumulative distribution function (CDF). Denoting $g(\Delta) = \phi(\Delta) - \Delta(1 - \Phi(\Delta))$ we see that $Dev_{S_j} = \sigma_j \, g(\Delta_j)$. With two bins Dev is a function from $[0,1]^2$ to $\mathbf{R}$ and $Dev(a,b) = \sigma_1 g(\Delta_1) + \sigma_2 g(\Delta_2)$ where the first bin has mean $a\mu$ and variance bV, the second bin has mean $(1-a)\mu$ and variance $(1-b)V$ and σ_j, Δ_j are defined as above.

Lemma 3. *Dev respects the symmetry, uni-modality and central saddle point properties.*

Proof. – <u>Symmetry:</u> Let us define $\sigma_1(b) = \sqrt{b}\,\sigma$, $\sigma_2(b) = \sqrt{1-b}\,\sigma$, $\Delta_1(a,b) = \frac{c_1 - a\mu}{\sigma_1(b)}$ and $\Delta_2(a,b) = \frac{c_2 - (1-a)\mu}{\sigma_2(b)}$. We know that $Dev(a,b) = \sigma_1(b)g(\Delta_1(a,b)) + \sigma_2(b)g(\Delta_2(a,b))$. To prove the symmetry $Dev(a,b) = Dev(1-a-\frac{c_2-c_1}{\mu}, 1-b)$, it is enough to show that the following four equations hold: $\sigma_1(b) = \sigma_2(1-b)$, $\sigma_2(b) = \sigma_1(1-b)$, $\Delta_1(a,b) = \Delta_2(1-a+\frac{c_1-c_2}{\mu}, 1-b)$ and $\Delta_2(a,b) = \Delta_1(1-a+\frac{c_1-c_2}{\mu}, 1-b)$.

Indeed, $\sigma_1(1-b) = \sqrt{1-b}\,\sigma = \sigma_2(b)$ and similarly $\sigma_2(1-b) = \sigma_1(b)$. Also, $\Delta_2(1-a-\frac{c_2-c_1}{\mu}, 1-b) = \frac{c_2 - (1-(1-a+\frac{c_1-c_2}{\mu}))\mu}{\sigma_2(1-b)}$. A similar check shows that $\Delta_1(1-a-\frac{c_2-c_1}{\mu}, 1-b) = \Delta_2(a,b)$. This proves the symmetry. We remark that for $c_1 = c_2$ this simply says we can switch the names of the first and second bin.

– <u>Uni-modality in a:</u> Calculations show that $\frac{\partial^2 Dev}{\partial a^2} = \mu^2 \left[\frac{\phi(\Delta_2)}{\sigma_2} + \frac{\phi(\Delta_1)}{\sigma_1} \right] \geq 0$. It follows that for any $0 < b < 1$, $Dev(a)$ is convex and has a unique minimum. The unique point $(m(b), b)$ on the valley is the one where $\Delta_1 = \Delta_2$.

– <u>Central saddle point:</u> We first explicitly determine what Dev restricted to the valley is as a function $D(b) = Dev(m(b), b)$ of b. As $Dev(a,b) = \sigma_1 g(\Delta_1) + \sigma_2 g(\Delta_2)$ and on the valley $\Delta_1 = \Delta_2$ we see that on the valley $Dev(a,b) = (\sigma_1 + \sigma_2)g(\Delta_1)$. However, $\sigma_1 + \sigma_2$ also simplifies to $\frac{c_1 - a\mu}{\Delta_1} + \frac{c_2 - (1-a)\mu}{\Delta_2} = \frac{c - \mu}{\Delta_1}$. Altogether, we conclude that on the valley $Dev(a,b) = (c - \mu)\frac{g(\Delta_1)}{\Delta_1}$ is a function of Δ_1 alone.

It is a straight forward calculation that $\frac{\partial Dev(\Delta_1)}{\partial \Delta_1} = -(c - \mu)\frac{\phi(\Delta_1)}{\Delta_1^2} < 0$. We will also show that $\frac{\partial \Delta_1}{\partial b}$ is negative when $b \leq \frac{1}{2}$ and positive when $b \geq \frac{1}{2}$. As $\frac{\partial D}{\partial b} = \frac{\partial Dev}{\partial \Delta_1} \cdot \frac{\partial \Delta_1}{\partial b}$, we see that $D(b)$ is increasing for $b \leq \frac{1}{2}$ and decreasing for $b \geq \frac{1}{2}$ as claimed.

To analyze $\frac{\partial \Delta_1}{\partial b}$ we write $\Delta_1 = \frac{e_1}{\sigma_1}$ and $\Delta_2 = \frac{e_2}{\sigma_2}$ where $e_1 = c_1 - a\mu$ is the spare capacity in bin 1 and $e_2 = c_2 - (1-a)\mu$ is the spare capacity in bin 2. We notice that $e = e_1 + e_2 = c - \mu$ the total spare capacity in the system. Now $\Delta_1 = \Delta_2$ implies $e_1\sigma_2 = e_2\sigma_1 = (e - e_1)\sigma_1$. Therefore, $e_1(\sigma_1 + \sigma_2) = e\sigma_1$ and $\Delta_1 = \frac{e}{\sigma_1 + \sigma_2} = \frac{c - \mu}{\sigma}\left(\frac{1}{\sqrt{b} + \sqrt{1-b}}\right)$ and notice that $\Delta = \frac{c - \mu}{\sigma}$ is independent of b. All that remains is to differentiate the function $\frac{1}{\sqrt{b} + \sqrt{1-b}}$.

We remark that we could simplify the proof by using Lagrange multipliers. However, since here it is easy to explicitly find Dev restricted to the valley we prefer the explicit solution. Later, we will not be able to explicitly find the restriction to the valley and we use instead Lagrange multipliers that solves the problem with an *implicit* description of the valley.

B Proving SP-MWOP Falls into Our Framework

Recall that
$$WOFP = \max_{j=1}^{k} \{1 - \Phi(\Delta_j)\}.$$

With two bins $WOFP$ is a function from $[0,1]^2$ to $\mathbf{R}$ and $WOFP(a,b) = \max\{1 - \Phi(\Delta_1), 1 - \Phi(\Delta_2)\}$ where the first bin has mean $a\mu$ and variance bV, the second bin has mean $(1-a)\mu$ and variance $(1-b)V$. σ_j, Δ_j were previously defined.

Lemma 4. *$WOFP$ respects the symmetry, uni-modality and central saddle point properties.*

Proof. – Symmetry: The same proof as in Appendix A shows $WOFP(a,b) = WOFP(1 - a - \frac{c_2 - c_1}{\mu}, 1 - b)$.
- Uni-modality in a: Fix b. Denote $OFP_1(a,b) = OFP_1(a\mu, bV) = 1 - \Phi(\Delta_1)$. It is a simple calculation that $\frac{\partial OFP_1}{\partial a}(a,b) = \frac{\mu}{\sqrt{b\sigma}} \cdot \phi(\Delta_1) > 0$. Similarly, if $OFP_2(a,b)$ denotes the overflow probability in the second bin when the first bin has total mean $a\mu$ and total variance bV, then $\frac{\partial OFP_2}{\partial a} = \frac{-\mu}{\sqrt{1-b\sigma}} \cdot \phi(\Delta_2) < 0$. Thus, OFP_1 is monotonically increasing in a and OFP_2 is monotonically decreasing in a, and therefore there is a unique minimum for $OFP(a,b)$ (when b is fixed and a is free) that is obtained when $OFP_1(a,b) = OFP_2(a,b)$, i.e., when $\Delta_1 = \Delta_2$.
- Central saddle point: We first explicitly determine what $WOFP$ restricted to the valley is as a function $D(b) = WOFP(m(b), b)$ of b. From before we know that on the valley $\Delta_1 = \Delta_2$. Therefore, following the same reasoning as in the SP-MED case,

$$\Delta_1(b) = \frac{c - \mu}{\sigma} \cdot \frac{1}{\sqrt{b} + \sqrt{1 - b}}.$$

It follows that $D(b)$ is monotonically decreasing in b for $b \leq \frac{1}{2}$ and increasing otherwise. The maximal point is obtained in the saddle point that is the center of the symmetry.

By Theorem 2 we know that the optimal fractional solution is obtained on the bottom sorted path. In fact, for SP-MWOP we can say a bit more:

Lemma 5. *The optimal fractional solution for SP-MWOP is the unique point that is the intersection of the valley and the bottom sorted path, and in this point $\Delta_1 = \Delta_2$.*

Proof. Let us assume by contradiction that the optimal point $P^* = (a^*, b^*)$ is not the point I which is the intersection point of the valley and the bottom sorted path. By Theorem 2, P^* is on the bottom sorted path. W.l.o.g. let us assume that P^* is left to the valley (the other case is similar). Since the valley is the curve defined by $\Delta_1 = \Delta_2$, it is easy to see that $\Delta_1(a^*, b^*) > \Delta_2(a^*, b^*)$

and therefore $WOFP(a^*, b^*) = 1 - \Phi(\Delta_2(a^*, b^*))$. Now, let us look at the point $P' = \frac{I+P^*}{2} = (a', b')$. P' is within the polygon confined by the bottom and upper sorted paths (by convexity) and is also left to the valley. Also, $a' > a^*$ and $b' \geq b^*$ and as before, $\Delta_1(a', b') > \Delta_2(a', b')$ and $WOFP(a', b') = 1 - \Phi(\Delta_2(a', b'))$. Moreover, Δ_2 is monotonically increasing in a and in b (i.e., $\frac{\partial \Delta_2}{\partial a}(a, b) > 0$ and $\frac{\partial \Delta_2}{\partial b}(a, b) > 0$), so $\Delta_2(a', b') > \Delta_2(a^*, b^*)$, and therefore $WOFP(a', b') < WOFP(a^*, b^*)$, in contradiction to the optimality assumption of the point P^*. Therefore, we must conclude that $P^* = I$.

C   Proving SP-MOP Falls into Our Framework

Recall that $OFP = 1 - \prod_{j=1}^{k}(1 - OFP_j)$ where $OFP_j = 1 - \Phi(\Delta_j)$. With two bins OFP is a function from $[0, 1]^2$ to $\mathbf{R}$ and $OFP(a, b) = 1 - \Phi(\Delta_1)\Phi(\Delta_2)$ where the first bin has mean $a\mu$ and variance bV, the second bin has mean $(1 - a)\mu$ and variance $(1 - b)V$ and σ_j, Δ_j were previously defined.

Lemma 6. *OFP respects the symmetry and uni-modality properties.*

Proof. – <u>Symmetry</u>: The same proof as in Appendix A shows $OFP(a, b) = OFP(1 - a - \frac{c_2 - c_1}{\mu}, 1 - b)$.
 – <u>Uni-modality in a</u>: Fix b. $\frac{\partial^2 OFP}{\partial a^2} = \mu^2[\frac{\Delta_1}{\sigma_1^2}\phi(\Delta_1)\Phi(\Delta_2) + \frac{\Delta_2}{\sigma_2^2}\phi(\Delta_2)\Phi(\Delta_1) + \frac{2}{\sigma_1 \sigma_2}\phi(\Delta_1)\phi(\Delta_2)]$. In particular $\frac{\partial^2 OFP}{\partial a^2} > 0$ and for every fixed b, $OFP(a, b)$ is convex over $a \in [0..1]$ and has a unique minimum $a = m(b)$.

Proving there exists a unique maximum over the valley is more challenging. We wish to find all extremum points of the cost function D (OFP in our case) over the valley $\{(m(b), b)\}$. Define $V(a, b) = a - m(b)$. Then we wish to maximize $D(a, b)$ subject to $V(a, b) = 0$. Before, we computed the restriction $D(b)$ of the cost function over the valley and found its extremum points. However, here we do not know how to explicitly find $D(b)$. Instead, we use Lagrange multipliers that allow working with the implicit form $V(a, b) = 0$ without explicitly finding $D(b)$. We prove a general result:

Lemma 7. *If a cost function D is differentiable twice over $[0, 1] \times [0, 1]$, then any extremum point Q of D over the valley must have zero gradient at Q, i.e., $\nabla(D)(Q) = 0$.*

Proof. Using Lagrange multipliers we find that at any extremum point Q of D over the valley,

$$\nabla(D)(Q) = \lambda \nabla V(Q). \qquad (3)$$

For some real value λ. However,

$$\nabla(D)(Q) = (\frac{\partial D}{\partial a}(Q), \frac{\partial D}{\partial b}(Q)) = (0, \frac{\partial D}{\partial b}(Q)),$$

because Q is on the valley and $\frac{\partial D}{\partial a}(Q) = 0$. As $V(a, b) = a - m(b)$, $\frac{\partial V}{\partial a}(Q) = 1$. We conclude that $\lambda = 0$. This implies that $\frac{\partial D}{\partial b}(Q) = 0$. Hence, $\nabla(D)(Q) = 0$.

Lemma 8. *OFP respects the central saddle point property.*

Proof. Let $Q = (a, b)$ be an extremum point of OFP over the valley. We look at the range $b \in [0..\frac{1}{2})$, $b \geq \frac{1}{2}$ is obtained by the symmetry. Then, by Lemma 7:

$$\phi(\Delta_1)\Phi(\Delta_2)\frac{\partial\Delta_1}{\partial a} = -\phi(\Delta_2)\Phi(\Delta_1)\frac{\partial\Delta_2}{\partial a} \text{ , and}$$

$$\phi(\Delta_1)\Phi(\Delta_2)\frac{\partial\Delta_1}{\partial b} = -\phi(\Delta_2)\Phi(\Delta_1)\frac{\partial\Delta_2}{\partial b}.$$

Dividing the two equations we get

$$\frac{\partial\Delta_1}{\partial a}\frac{\partial\Delta_2}{\partial b} = \frac{\partial\Delta_2}{\partial a}\frac{\partial\Delta_1}{\partial b}.$$

Plugging the partial derivatives of Δ_i by a and b, we get the equation

$$\frac{\Delta_1}{\Delta_2} = \sqrt{\frac{b}{1-b}}.$$

As $b \leq \frac{1}{2}$, $b < 1 - b$ and we conclude that at Q $\Delta_1 < \Delta_2$. However, using the log-concavity of the normal c.d.f function Φ we prove that:

Lemma 9. $\frac{\partial OFP}{\partial a} = 0$ *at a point* $Q = (a, b)$ *with* $b \leq \frac{1}{2}$ *implies* $\Delta_1 \geq \Delta_2$.

Proof. The condition $\frac{\partial OFP}{\partial a} = 0$ is equivalent to

$$\frac{\phi(\Delta_1)}{\Phi(\Delta_1)} = \frac{\sigma_1}{\sigma_2} \cdot \frac{\phi(\Delta_2)}{\Phi(\Delta_2)}$$

As $b < \frac{1}{2}$, $b < 1 - b$ and $\sigma_1 < \sigma_2$. Hence,

$$\frac{\phi(\Delta_1)}{\Phi(\Delta_1)} < \frac{\phi(\Delta_2)}{\Phi(\Delta_2)}.$$

Denote $h(\Delta) = \frac{\phi(\Delta)}{\Phi(\Delta)}$. We will prove that h is monotone decreasing, and this implies that $\Delta_1 > \Delta_2$.

To see that h is monotone decreasing define $H(\Delta) = \ln(\Phi(\Delta))$. Then $h = H'$. Therefore, $h' = H''$. However, Φ is log-concave, hence $H'' < 0$. We conclude that $h' < 0$ and h is monotone decreasing.

Together, this implies that the only extremum point of OFP over the valley is at $b = \frac{1}{2}$. However, at $b = 0$, the best is to fill the largest bin to full capacity with variance 0, and thus, $OFP(m(0), 0) = 1 - \Phi(\Delta)$ where $\Delta = \frac{c-\mu}{\sigma}$. On the other hand, at $b = \frac{1}{2}$, $OFP(a = m(\frac{1}{2}), \frac{1}{2}) = 1 - \Phi(\frac{c_1-a\mu}{\sqrt{\frac{1}{2}}\sigma})\Phi(\frac{c_2-(1-a)\mu}{\sqrt{\frac{1}{2}}\sigma})$. As $(c_1 - a\mu) + (c_2 - (1-a)\mu) = c - \mu$, either $c_1 - a\mu$ or $c_2 - (1-a)\mu$ is at most $\frac{c-\mu}{2}$ and therefore $\Phi(\sqrt{2}\frac{c_1-a\mu}{\sigma})\Phi(\sqrt{2}\frac{c_2-(1-a)\mu}{\sigma}) \leq \Phi(\sqrt{2}\frac{c-\mu}{2\sigma}) = \Phi(\frac{c-\mu}{\sqrt{2}\sigma}) \leq \Phi(\frac{c-\mu}{\sigma})$. We conclude that $OFP(a, \frac{1}{2}) \geq OFP(m(0), 0)$ and there is a unique maximum point on the valley and it is obtained at $b = \frac{1}{2}$.

D   Error Induced by the Reduction to the Normal Distribution

The error in our algorithm stems from two different parts:

- The error induced by the reduction to the normal case, and
- The error the algorithm has on the normal distribution, mainly induced because the algorithm outputs an integral solution rather than the optimal fractional solution.

We analyze separately each kind of error and in this section we analyze the error induced by the reduction to the normal case. For SP-MED we gave a complete analysis of the error in Proposition 1. The analogous (and simpler) Proposition for SP-MWOP is:

Proposition 3 *(SP-MWOP). Given n independent random variables $X = (X^{(1)}, \ldots, X^{(n)})$ with mean $\mu^{(i)}$, variance $V^{(i)} = \mathbb{E}(|X^{(i)} - \mu^{(i)}|^2)$ and $\rho^{(i)} = \mathbb{E}(|X^{(i)} - \mu^{(i)}|^3)$ and a partition $S = \{S_1, \ldots, S_k\}$,*

$$|D_X(S) - D_N(S)| \leq C_0 \cdot \psi_0^{max}(S),$$

where C_0 is the constant defined in Theorem 5,

$$\psi_0^j(S_j) = \frac{\sum_{i \in S_j} \rho^{(i)}}{(\sum_{i \in S_j} V^{(i)})^{\frac{3}{2}}},$$

and $\psi_0^{max}(S) = \max_{j=1}^k \psi_0^j(S_j)$.

Proof. Let $X_j = \sum_{i \in S_j} X^{(i)}$, $N_j = N(\sum_{i \in S_j} \mu^{(i)}, \sum_{i \in S_j} V^{(i)})$ and F_{X_j}, F_{N_j} be their cumulative distribution functions. Then, for every j,

$$|\Pr_X(OF_j) - \Pr_N(OF_j)| = |1 - F_{X_j}(c_j) - (1 - F_{N_j}(c_j))|$$

$$= |F_{X_j}(c_j) - F_{N_j}(c_j)| \leq C_0 \cdot \psi_0^j(S_j),$$

where the inequality is by Theorem 5. Let j' be the bin with maximum overflow probability under X, and j'' be the bin with maximum overflow probability under N. Clearly,

$$\Pr_X(OF_{j'}) \geq \Pr_X(OF_{j''})$$

$$\geq \Pr_N(OF_{j''}) - C_0 \cdot \psi_0^{j''}(S_{j''})$$

$$\geq \Pr_N(OF_{j'}) - C_0 \cdot \psi_0^{j''}(S_{j''})$$

$$\geq \Pr_X(OF_{j'}) - C_0 \cdot \psi_0^{j'}(S_{j'}) - C_0 \cdot \psi_0^{j''}(S_{j''})$$

Therefore,

$$\Pr_N(OF_{j''}) - \Pr_X(OF_{j'}) \leq C_0 \cdot \psi_0^{j''}(S_{j''}), \text{ and,}$$

$$\Pr_N(OF_{j''}) - \Pr_X(OF_{j'}) \geq -C_0 \cdot \psi_0^{j'}(S_{j'})$$

and hence,

$$|D_X(S) - D_N(S)| = |\Pr_X(OF_{j'}) - \Pr_N(OF_{j''})| \leq C_0 \cdot \psi_0^{max}(S)$$

A similar argument works for SP-MOP using

$$D_X(S) = 1 - \Pi_{j=1}^k (1 - \Pr_X(OF_j)),$$

$$D_N(S) = 1 - \Pi_{j=1}^k (1 - \Pr_N(OF_j))$$

and,

$$|D_X(S) - D_N(S)| = |\Pi_{j=1}^k (1 - \Pr_X(OF_j)) - \Pi_{j=1}^k (1 - \Pr_N(OF_j))|$$

$$\leq \sum_{j=1}^k |\Pr_X(OF_j) - \Pr_N(OF_j)| \leq C_0 \cdot \sum_{j=1}^k \psi_0^j.$$

E   Error Induced by Outputting an Integral Solution

Here we need to show that rounding the fractional solution to integral in the Normal case does not induce much error. For that we need to assume that the system has some spare capacity and that no input service is too *dominant*. We define two parameters:

- Spare capacity: We define a new system constant, *relative spare capacity*, denoted by α where

$$\alpha = \frac{c - \mu}{\mu},$$

i.e., it expresses the spare capacity as a fraction of the total mean. We assume that the system has some constant (possibly small) relative spare capacity.
- No dominant service: As before, we represent service i with the point $P^{(i)} = (a^{(i)}, b^{(i)})$ and $P^{(1)} + P^{(2)} + \ldots + P^{(n)} = (1, 1)$. Thus, $\sum_i |P^{(i)}| \geq |(1,1)| = \sqrt{2}$ (by the triangle inequality) and $\sum_i |P^{(i)}| \leq 2$ (because the length of the longest increasing path from $(0,0)$ to $(1,1)$, is obtained by the path going from $(0,0)$ to $(1,0)$ and then to $(1,1)$). Hence, the average length of an input point $P^{(i)}$ is somewhere between $\frac{\sqrt{2}}{n}$ and $\frac{2}{n}$. Our assumption states that no element takes more than L times its "fair" share, i.e., that for every i, $|P^{(i)}| \leq \frac{L}{n}$.

Also, we only consider solutions where each bin is allocated services with total mean not exceeding its capacity. Equivalently, we only consider solutions where $\Delta_j \geq 0$ for every $1 \leq j \leq k$. We will later see that under these conditions the sorting algorithm solves all three cost functions with a small error going fast to zero with n. We prove:

Theorem 6. *Let OPT_f be the fractional optimal solution. If D is differentiable, the difference between the cost on the integral point found by the sorting algorithm and the cost on the optimal integral (or fractional) point is at most* $\min\{|\nabla D(\xi_1)|, |\nabla D(\xi_2)|\}\frac{L}{n}$, *where $\xi_1 \in [O_1, OPT_f]$, $\xi_2 \in [OPT_f, O_2]$ and O_1 and O_2 are the two points on the bottom sorted path between which OPT_f lies.*

Proof. Suppose we run the sorting algorithm on some input. Let OPT_{int} be the integral optimal solution, OPT_f the fractional optimal solution and OPT_{sort} the integral point the sorting algorithm finds on the bottom sorted path. We wish to bound $D(OPT_{sort}) - D(OPT_{int})$ and clearly it is at most $D(OPT_{sort}) - D(OPT_f)$. We now look at the two points O_1 and O_2 on the bottom sorted path between which OPT_f lies (and notice that as far as we know it is possible that OPT_{sort} is none of these points). Since $D(OPT_f) \leq D(OPT_{sort}) \leq D(O_1)$ and $D(OPT_f) \leq D(OPT_{sort}) \leq D(O_2)$ the error the sorting algorithm makes is at most

$$\min\{D(O_1) - D(OPT_f), D(O_2) - D(OPT_f)\}.$$

We now apply the mean value theorem and use our assumption that for every i, $|P^{(i)}| \leq \frac{L}{n}$.

We remark that in fact the proof shows something stronger: the cost of any (not necessarily optimal) fractional solution on the bottom sorted path, is close to the cost of the integral point to the left or to the right of it on the bottom sorted path. We now specialize Theorem 6 for SP-MED and SP-MWOP.

E.1   SP-MED

Lemma 10. *The difference between the expected deviation in the integral point found by the sorting algorithm and the optimal integral (or fractional) point for SP-MED is at most* $\frac{1}{\sqrt{2\pi e}} \cdot \frac{1}{\alpha} \cdot \frac{L}{n} \cdot \mu$. *In particular, when $L = o(n)$ and α is a constant, the error is $o(\mu)$.*

Proof. We know from Theorem 6 that the difference is at most

$$\min\{|\nabla D(\xi_1)|, |\nabla D(\xi_2)|\}\frac{L}{n},$$

where $\xi_1 \in [O_1, OPT_f]$, $\xi_2 \in [OPT_f, O_2]$ and O_1 and O_2 are the two points on the bottom sorted path between which OPT_f lies. Plugging the partial derivatives, we see that

$$|\nabla(\sigma_2 g)(\Delta_2)| \leq |\mu(1 - \Phi(\Delta_2))| + |\frac{\sigma}{2\sqrt{1-b}}\phi(\Delta_2)|$$

$$\leq \mu + \frac{\sigma}{2\sqrt{1-b}}\phi(\Delta_2).$$

Moreover, $\frac{\sigma}{2\sqrt{1-b}}\phi(\Delta_2) = \frac{\sigma}{2\sqrt{1-b}}\frac{1}{\Delta_2}\Delta_2\phi(\Delta_2)$ and a simple calculation shows that the function $\Delta\phi(\Delta)$ maximizes at $\Delta = 1$ with value at most $\frac{1}{\sqrt{2\pi e}}$. By our assumption that $\Delta_j \geq 0$ for every j, we get that

$$\frac{\sigma}{2\sqrt{1-b}}\phi(\Delta_2) \leq \frac{\sigma}{2\sqrt{1-b}}\frac{\sigma\sqrt{1-b}}{c_2 - (1-a)\mu}\frac{1}{\sqrt{2\pi e}}$$

$$\leq \frac{V}{2\sqrt{2\pi e}}\frac{1}{c_2 - (1-a)\mu}.$$

Applying the same argument on O_2 shows the error can also be bounded by $\frac{V}{2\sqrt{2\pi e}}\frac{1}{c_1 - a\mu}$.

However, $(c_1 - a\mu) + (c_2 - (1-a)\mu) = c - \mu$ which is the total spare capacity, and at least one of the bins takes spare capacity that is at least half of that, namely $\frac{c-\mu}{2}$. Since the error is bounded by either term, we can choose the one where the spare capacity is at least $\frac{c-\mu}{2}$ and we therefore see that the error is at most $\frac{V}{2\sqrt{2\pi e}}\frac{2}{c-\mu}$. Since we assume $c - \mu \geq \alpha\mu$ for some constant $\alpha > 0$, the error is at most $\frac{V}{\sqrt{2\pi e}}\frac{1}{\alpha\mu}$. As we assume $V \leq \mu^2$, $\frac{V}{\mu} \leq \mu$ which completes the proof.

This shows the approximation factor goes to 1 and linearly (in the number of services) fast. Thus, from a practical point of view the theorem is very satisfying.

E.2 SP-MWOP

Lemma 11. *The difference between minimal worst overflow probability in the integral point found by the sorting algorithm and the optimal integral (or fractional) point for SP-MWOP is at most $O(\frac{L}{\alpha n})$. In particular, when $L = {}_o(n)$ and α is a constant, the difference is ${}_o(1)$.*

Proof. We know from Theorem 6 that the difference is at most

$$\min\{|\nabla D(\xi_1)|, |\nabla D(\xi_2)|\}\frac{L}{n},$$

where $\xi_1 = (a_1, b_1) \in [O_1, OPT_f]$, $\xi_2 = (a_2, b_2) \in [OPT_f, O_2]$ and O_1 and O_2 are the two points on the bottom sorted path between which OPT_f lies, and notice that even though $WOFP$ is not differentiable when $\Delta_1 = \Delta_2$, it is differentiable everywhere else. We plug the partial derivatives and also replace $\frac{\phi(\Delta_2)}{\sigma_2}$ with $\frac{\Delta_2\phi(\Delta_2)}{c_2 - (1-a)\mu}$ and similarly for the other term. We get:

$$\min\left\{|\Delta_2\phi(\Delta_2)| \cdot |(\frac{\mu}{c_2 - (1-a_1)\mu}, \frac{1}{2(1-b_1)})|, |\Delta_1\phi(\Delta_1)| \cdot |(\frac{\mu}{c_1 - a_2\mu}, \frac{1}{2b_2})|\right\}\frac{L}{n}$$

The term $\Delta\phi(\Delta)$ maximizes at $\Delta = 1$ with value at most $\frac{1}{\sqrt{2\pi e}}$. Also, $(c_1 - a_2\mu) + (c_2 - (1-a_1)\mu) = c - \mu - (a_2 - a_1)\mu \geq c - \mu\frac{L}{n} \geq \frac{\alpha}{2}\mu$, where α is the total space capacity, and a constant by our assumption. Hence, at least one of the terms $\frac{\mu}{c_2-(1-a_1)\mu}, \frac{\mu}{c_1-a_2\mu}$ is at most $\frac{4}{\alpha}$. Also, for that term, the spare capacity is maximal, and therefore it takes at least half of the variance. Altogether, the difference is at most $O(\frac{L}{\alpha n})$ which completes the proof.

F   Unbalancing Bin Capacities Is Always Better

Suppose we are given a capacity budget c and we have the freedom to choose capacities c_1, c_2 that sum up to c for two bins. Which choice is the best? Offhand, it is possible that for each input there is a different choice of c_1 and c_2 that minimizes the cost. In contrast, we show that for the three cost functions we consider in this paper, the minimum cost always decreases as the difference $c_2 - c_1$ increases.

Lemma 12. *Given a capacity budget c and either SP-MED, SP-MWOP or SP-MOP cost function, the minimum cost decreases as c_2-c_1 increases. In particular the best choice is having a single bin with capacity c and the worst choice is splitting the capacities evenly between the two bins.*

Proof. Recall that $\Delta_1(a,b) = \frac{c_1-a\mu}{\sigma\sqrt{b}}$ and $\Delta_2(a,b) = \frac{c_2-(1-a)\mu}{\sigma\sqrt{1-b}}$. Therefore, if we reduce c_1 by $\tilde{c}$ and increase c_2 by $\tilde{c}$, we get

$$\tilde{\Delta}_1(a,b) \overset{def}{=} \frac{c_1-\tilde{c}-a\mu}{\sigma\sqrt{b}} = \frac{c_1-(a+\frac{\tilde{c}}{\mu})\mu}{\sigma\sqrt{b}} = \Delta_1(a+\frac{\tilde{c}}{\mu},b).$$

Similarly, $\tilde{\Delta}_2(a,b) = \Delta_2(a+\frac{\tilde{c}}{\mu},b)$.

Let $Dev_{c_1,c_2}(a,b)$ denote the expected deviation with bin capacities c_1,c_2, $WOFP_{c_1,c_2}(a,b)$ denote the worst overflow probability with bin capacities c_1,c_2 and $OFP_{c_1,c_2}(a,b)$ denote the overflow probability with bin capacities c_1,c_2. As

$$Dev(a,b) = \sigma_1(b)g(\Delta_1(a,b)) + \sigma_2(b)g(\Delta_2(a,b)),$$

$$WOFP(a,b) = \max\{1 - \Phi(\Delta_1), 1 - \Phi(\Delta_2)\} \text{ and,}$$

$$OFP(a,b) = 1 - \Phi(\Delta_1)\Phi(\Delta_2), \text{ we see that:}$$

$$Dev_{c_1-\tilde{c},c_2+\tilde{c}}(a,b) = Dev_{c_1,c_2}(a+\frac{\tilde{c}}{\mu},b),$$

$$WOFP_{c_1-\tilde{c},c_2+\tilde{c}}(a,b) = WOFP_{c_1,c_2}(a+\frac{\tilde{c}}{\mu},b),$$

$$OFP_{c_1-\tilde{c},c_2+\tilde{c}}(a,b) = OFP_{c_1,c_2}(a+\frac{\tilde{c}}{\mu},b) \text{ and,}$$

i.e., each cost graph is shifted left by $\frac{\tilde{c}}{\mu}$.

Notice that the bottom sorted path does not depend on the bin capacities and is the same for every value of c_1 and c_2 we choose. Let (a,b) be the optimal fractional solution for bin capacities c_1,c_2. We know that (a,b) is on the bottom sorted path. Let $\tilde{a} = a - \frac{\tilde{c}}{\mu}$. We saw that the cost function

$D \in \{Dev, WOFP, OFP\}$ satisfies $D_{c_1-\tilde{c},c_2+\tilde{c}}(\tilde{a},b) = D_{c_1,c_2}(a,b)$. The point $(\tilde{a},b)$ lies to the left of the bottom sorted path and therefore above it. As the optimal solution for bin capacities $c_1 - \tilde{c}, c_2 + \tilde{c}$ is also on the bottom sorted path and is strictly better than any internal point, we conclude that the expected deviation for bin capacities $c_1 - \tilde{c}, c_2 + \tilde{c}$ is strictly smaller than the expected deviation for bin capacities c_1, c_2.

An immediate corollary is the trivial fact that putting all the capacity budget in one bin is best. Obviously, this is not always possible nor desirable, but if there is tolerance in each bin capacity, we recommend minimizing the number of bins.

Our simulation results, both on synthetic normally distributed data and on real independent data, also clearly show this phenomenon. Figure 9 shows the cost of the sorting algorithm for the three cost functions as a function of $\frac{c}{\mu}$, for $\frac{c_1}{c} \in \{0.1, 0.2, 0.3, 0.4, 0.5\}$ and synthetic normally distributed data. We can clearly see that the cost decreases as $\frac{c_1}{c}$ decreases in both data sets. The results for real independent data are very similar and we omit them due to lack of space.

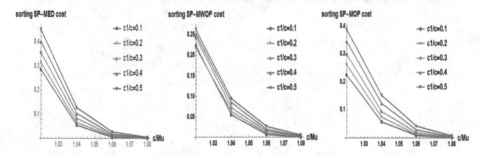

Fig. 9. Average two bins cost of the sorting algorithm for several values of $\frac{c_1}{c}$ and **synthetic normally distributed data**. Three cost functions that are considered: SP-MED, SP-MWOP and SP-MOP. The x axis measures $\frac{c}{\mu}$.

References

1. Kleinberg, J., Rabani, Y., Tardos, É.: Allocating bandwidth for bursty connections. SIAM J. Comput. **30**(1), 191–217 (2000)
2. Goel, A., Indyk, P.: Stochastic load balancing and related problems. In: IEEE FOCS 1999, pp. 579–586 (1999)
3. Wang, M., Meng, X., Zhang, L.: Consolidating virtual machines with dynamic bandwidth demand in data centers. In: IEEE INFOCOM 2011, pp. 71–75 (2011)
4. Breitgand, D., Epstein, A.: Improving consolidation of virtual machines with risk-aware bandwidth oversubscription in compute clouds. In: IEEE INFOCOM 2012, pp. 2861–2865 (2012)

5. Nikolova, E., Kelner, J.A., Brand, M., Mitzenmacher, M.: Stochastic shortest paths via Quasi-convex maximization. In: Azar, Y., Erlebach, T. (eds.) ESA 2006. LNCS, vol. 4168, pp. 552–563. Springer, Heidelberg (2006). https://doi.org/10.1007/11841036_50
6. Nikolova, E.: Approximation algorithms for offline risk-averse combinatorial optimization. Approximation, Randomization, and Combinatorial Optimization, pp. 338–351. Algorithms and Techniques (2010)
7. Shabtai, G., Raz, D., Shavitt, Y.: Risk aware stochastic placement of cloud services: the multiple data center case. In: ALGOCLOUD 2017 (2017)
8. Esseen, C.G.: A moment inequality with an application to the central limit theorem. Scand. Actuarial J. **1956**(2), 160–170 (1956)
9. Shevtsova, I.: An improvement of convergence rate estimates in the Lyapunov theorem. Dokl. Math. **82**, 862–864 (2010)
10. Ibragimov, I.A., Linnik, Y.V.: Independent and Stationary Sequences of Random Variables. Wolters-Noordhoff, Groningen (1971)
11. O'Donnell, R.: Analysis of Boolean Functions. Cambridge University Press, Cambridge (2014)
12. Michael, R.G., David, S.J.: Computers and Intractability: A Guide to the Theory of NP-Completeness. W.H. Freeman, New York (1979)

Towards an Algebraic Cost Model for Graph Operators

Alexander Singh[✉] ⓘ and Dimitrios Tsoumakos

Department of Informatics, Ionian University, Corfu, Greece
{p13sing,dtsouma}@ionio.gr

Abstract. Graph Analytics has been gaining an increasing amount of attention in recent years. This has given rise to the development of numerous graph processing and storage engines, each featuring different models in computation, storage and execution as well as performance. Multi-Engine Analytics present a solution towards adaptive, cost-based complex workflow scheduling to the best available underlying technology. To achieve this in the Graph Analytics case, detailed and accurate cost models for the various runtimes and operators must be defined and exported, such that intelligent planning can take place. In this work, we take a first step towards defining a cost model for graph-based operators based on an algebra and its primitives. We evaluate its accuracy over a state of the art graph database and discuss its advantages and shortcomings.

1 Introduction

In recent years, we observe an increasing interest in Graph Data Analytics. Initially driven by the surge in social graph data and analysis, graph analytics can be utilized to effectively (and intuitively in many cases) tackle multiple tasks in bioinformatics, social community analysis, traffic optimization in IoT, optimization/robustness in power grids and large networks, RDF data, etc. In graph analytics, the primitives of a data graph G, i.e., its vertices and edges are mapped to problem entities and the respective relationships between them.

A continuously expanding set of approaches and tools have emerged, in order to assist in efficient storage and computation of various algorithms over big data graphs ([1–3,9,11], etc.). These systems differ in one or more aspects, with the most important being their storage and computation (or execution) model [16]. However, one size does not fit all: No single execution model is suitable for all types of tasks and no single data model is suitable for all types of data. Modern workflows consist of series of diverse operators over heterogeneous data sources and formats. *Multi-Engine Analytics* has been proposed as a solution that can optimize for this complexity and is gaining ground ever since (e.g., [6,7,12]).

One of the most critical challenges in such a multi-engine environment is the design and creation of a *meta-scheduler* that automatically allocates tasks to the right engine(s) according to multiple criteria, deploys and runs them

© Springer International Publishing AG, part of Springer Nature 2018
D. Alistarh et al. (Eds.): ALGOCLOUD 2017, LNCS 10739, pp. 89–105, 2018.
https://doi.org/10.1007/978-3-319-74875-7_6

without manual intervention. For such a meta-scheduler to function properly, accurate *cost models* of the operators utilized by the underlying platforms or datastores must be defined or exported. The cost/performance tradeoffs will be then evaluated by the scheduler in order to adaptively "mix-and-match" operators to different engines in order to achieve a user-defined performance function (relating to time, cost, accuracy, etc.). Yet, while cost models have been thoroughly studied in traditional systems (e.g., RDBMSs or traditional big data platforms like Hadoop), this is not the case for Graph Analytics runtimes.

In this work, we take a first step towards this direction. Specifically, using the algebra defined in [10], we describe the decomposition of common graph operations down to primitive operators. We then devise a cost model and empirically evaluate it using a state of the art graph database [3] loaded with varying sizes and types of directed graphs and queries based on the presented decompositions. We assess the strengths and weaknesses of the resulting cost model and identify factors that contribute to its accuracy. Defining an engine-based cost model, that still originates from a general algebra, is a decisive step towards multi-engine cost-based optimization of graph analytics workflows.

2   Related Work

Inspired by the surging interest in graph databases, a number of works that focus on foundations of graphs and graph-databases, RDF and triple-stores have appeared.

An algebra for RDF Graphs was presented in [14], focused on querying large-scale, distributed triple-stores on shared-nothing clusters. The authors present the algebra as well as denotational semantics for it, while additionally presenting some preliminary experimental results. Another algebraic framework, also focusing on RDF data querying, was presented in [8]. This work focuses on defining an algebra to be used in the formal specification of an RDF query language. It is presented as a set of extraction, loop, construction and other operators, focusing on both the constructive and the extraction-related aspects of an RDF query language.

A SPARQL algebra to be used as the foundation for optimizing SPARQL-queries was presented in [15]. The authors define set-based semantics for SPARQL via a set algebra, further providing algebraic equivalence rules to be used for optimization purposes. They also identified fragments of SPARQL together with their complexity classes. Another SPARQL-related work was presented in [5], focusing on the transformations from SPARQL to the more traditional relational algebra framework. The author discusses this transformation framework, the mismatch between SPARQL and relational semantics and additionally outlines an SQL-based translation.

An algebra and related equivalence rules on transformations of graph patterns in Neo4j's property graph model is presented in [10]. A set of operators for retrieval and selection of edges and nodes is discussed. These operators result in the output of *graph relations*. Additionally, a set of equivalence rules is presented to aid in query optimization. A demonstration shows how these equivalences can be used to algebraically transform Cypher queries at the logical level, and a performance example is given, in terms of database hits, comparing the query evaluation plan found by Neo4j to the one given by the equivalences.

A presentation of a foundational framework for RDF databases is made by Arenas et al. [4], which includes a sound-and-complete deductive system focused on RDF-graph entailment. The work also includes discussions of algebraic syntax and compositional semantics for SPARQL, its expressive power, complexity considerations of evaluating various fragments of it and the optimization of SPARQL queries. A discussion on the issue of RDF-based queries in an RDFS framework is also included and an extension of SPARQL with navigational capabilities is presented.

3   Algebraic Framework

3.1   Data Model

The data model in this work is based on the notions of property graphs and graph relations (see [10]), defined as follows:

Definition 1. *Property Graph*
Let $G = (V, E, \Sigma_v, \Sigma_e, A_v, A_e, \lambda, L_v, L_e)$ be a property graph, where V is a set of nodes, E is a set of edges, Σ_v is a set of node labels, Σ_e is a set of edge labels. We also have that A_v is a set of node properties, while A_e is a set of edge properties. Let D be a set of atomic domains, then a property $\alpha_i \in A_v$ is a function $\alpha_i : V \to D_i \cup \{\epsilon\}$ assigning a property value from a domain $D_i \in D$ to a node $v \in V$, if v has property a_i - otherwise $a_i(v)$ returns ϵ. Accordingly, a property $\alpha_j \in A_e$ is a function $\alpha_j : E \to D_j \cup \{\epsilon\}$ which assigns a property value from a domain D_j to an edge $e \in E$, if e has property α_j, else $a_j(e) = \epsilon$. Finally, $\lambda : E \to V \times V$ is a function that assigns nodes to edges, $L_v : V \to \Sigma_v$ is a function that assigns labels to nodes, and $L_e : E \to \Sigma_e$ is a function assigning labels to edges.

Definition 2. *Graph Relation*
Let G be a property graph. Then, a relation R is a graph relation if the following is true:

$$\forall A \in attr(R) : dom(A) = V \cup E \qquad (3.1.1)$$

where $attr(R)$ is the set of attributes of R (i.e., columns) and $dom(A)$ is the domain of attribute A.

3.2   Base Operators

Our algebraic framework is built on the definition of two primitive operators, *getNodes* and *expand* [10]. These two operands can then be composed in order to produce higher-level graph operations. We proceed with their description, assuming a static graph G to operate on.

Definition 3. *getNodes Operator*
Consider a property graph G. The getNodes operator (denoted by ◯) takes as argument a label x and outputs a graph relation with a single attribute x containing all the nodes of G:

$$val(◯_x) = V, \tag{3.2.1}$$

$$sch(◯_x) = \langle x \rangle \tag{3.2.2}$$

where $val(R)$ is the set of tuples in R and $sch(R)$ is the schema of R.

Definition 4. *expand Operator*
Consider a property graph G. The expand operator, which takes as inputs a relation R, an attribute $x \in attr(R)$ and a new attribute label y, works by expanding the graph relation R to include the immediate neighbors of nodes under x reachable by an ingoing or outgoing edge. It does so by adding a new column y to R, containing the nodes than can be reached by an ingoing or outgoing edge (see below) from nodes of x. It comes in two variants, an expandIn (denoted by $\downarrow_x^y$) operator that selects neighbors that are reachable by ingoing edges, and an expandOut (denoted by $\uparrow_x^y$) which selects based on outgoing ones respectively:

$$val(\uparrow_x^y (R)) = \{\langle t, e, v \rangle | t \in val(R) \wedge e \in E \wedge \lambda(e) = (t.x, v)\} \tag{3.2.3}$$

$$sch(\uparrow_x^y (R)) = sch(R) \| \langle xy, y \rangle \tag{3.2.4}$$

where $val(R)$ is the set of tuples in R, $sch(R)$ is the schema of R, and $\|$ denotes concatenation of schema tuples. The semantics of $\downarrow_x^y$ can be defined in much the same way:

$$val(\downarrow_x^y (R)) = \{\langle t, e, v \rangle | t \in val(R) \wedge e \in E \wedge \lambda(e) = (v, t.x)\} \tag{3.2.5}$$

$$sch(\downarrow_x^y (R)) = sch(R) \| \langle yx, y \rangle \tag{3.2.6}$$

Apart from the aforementioned operators, we will also utilize the traditional relational algebra operators: select (σ), project (π) and join ($\bowtie$), which operate on graph relations with similar semantics to the traditional sense. As an example of the definitions above, consider the example of applying $\uparrow_x^y (◯_x)$ on a graph G, as shown in Fig. 1.

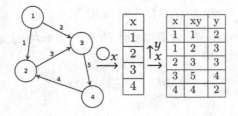

Fig. 1. Example of operations on graph

3.3   Cost Model

We now define space and time costs for the $getNodes$ and $expandIn/expandOut$ operators. We note that the cost model is implementation-specific with multiple practical factors such as algorithmic implementation, the underlying data structures, relational operator implementation, etc. having an effect on it. For simplicity, we consistently assume bare-bones implementations, i.e., graphs implemented as lists V, E containing nodes and edges (as node-tuples) respectively, joins using naive loop algorithms, etc. This assumption does not compromise our methodology, as factors can be modified and be "plugged in" to match the implementation at hand. We measure cost in terms of both (predicted) space/row count of the relations produced by each operator and (predicted) time/primitive operations taken to compute each query. The cost of the $getNodes$ operator only scales with respect to the number of nodes present in our graph:

$$SpaceCost(\bigcirc_{label}) = O(|V|) \tag{3.3.1}$$

$$OpCost(\bigcirc_{label}) = O(|V|) \tag{3.3.2}$$

The costs of the $expandIn$ and $expandOut$ operators scale with respect to the size of the input graph relation, as well as the average number of ingoing/outgoing edges per node of our graph respectively:

$$SpaceCost(\uparrow_{sourceLabel}^{targetLabel} R) = O(d_{out} \cdot SpaceCost(R))$$
$$SpaceCost(\downarrow_{targetLabel}^{sourceLabel} R) = O(d_{in} \cdot SpaceCost(R)) \tag{3.3.3}$$

$$OpCost(\uparrow_{sourceLabel}^{targetLabel} R) = O((|E| \cdot SpaceCost(R)) + OpCost(R))$$
$$OpCost(\downarrow_{targetLabel}^{sourceLabel} R) = O((|E| \cdot SpaceCost(R)) + OpCost(R)) \tag{3.3.4}$$

where R is a series of operations resulting in a graph relation, and d_{in} and d_{out} is the average number of ingoing and ougoing edges per node in G, the property graph we operate on.

Note that computing the cost for compositions of operators is done via recursion. For instance:

$$SpaceCost(\uparrow_x^y (\bigcirc_x)) = (d_{out} \cdot SpaceCost(\bigcirc_x)) = d_{out} \cdot |V| \tag{3.3.5}$$

$$OpCost(\uparrow_x^y (\bigcirc_x)) = O(|E| \cdot SpaceCost(\bigcirc_x) + OpCost(\bigcirc_x) = O((|E| \cdot |V|) + |V|) \tag{3.3.6}$$

4   Graph Operator Decomposition

In the following subsections we describe decompositions of frequently-used graph operations using the operators defined above.

4.1 Finding Cycles

Consider the task of querying a graph database for cycles. This is a popular operator found in many use cases (e.g., SPARQL query processing [13]). Expressing this query in terms of the base operators depends on the edge orientations we want to consider. For example, if we are to consider squares (cycles of size 4) formed with outgoing edges, the query can be expressed as seen in the left subfigure of Fig. 2. Similarly, we can make use of *expandIn* operators for the ingoing case, or even mix the two operators to find cycles of mixed, but defined beforehand, edge orientations.

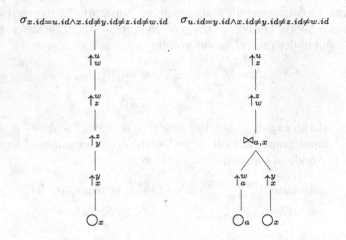

Fig. 2. Two alternative decompositions of the square query.

Another way to translate the above square query can be seen in the right subfigure of Fig. 2. Note that both queries will return graph relations that potentially contain rows that describe the same cycle – each one traversing a cycle's nodes in a different order.

We can now compute the costs of these two queries. We know that both queries will return the same results after the final selection, namely all the squares found in our graph. We also cannot compute the space costs for such a selection without knowing more about the number of squares in our graph. Thus, we only consider the costs up to that selection. For the first query, the cost is[1]:

$$SpaceCost(\uparrow_w^u \circ \uparrow_z^w \circ \uparrow_y^z \circ \uparrow_x^y (\bigcirc_x)) = O(d_{out}^4 \cdot |V|) \qquad (4.1.1)$$

$$OpCost(\sigma_K \circ \uparrow_w^u \circ \uparrow_z^w \circ \uparrow_y^z \circ \uparrow_x^y (\bigcirc_x)) = O((|E| \cdot (d_{out}^3 \cdot |V|)) + (|E| \cdot (d_{out}^2 \cdot |V|))$$
$$+ (|E| \cdot (d_{out} \cdot |V|)) + (|E| \cdot |V|) + |V|) \qquad (4.1.2)$$

---

[1] We use the notation ∘ to denote composition of functions.

where $K \rightarrow u.id = y.id \wedge x.id \neq y.id \neq z.id \neq w.id$. For the second query, we begin by observing that the two inputs of the join are essentially the same relation. From that we can deduce that the resulting relation has the same size as the first input multiplied by d_{out} (since for each row in the first argument, we'll get d_{out} rows in the joined relation), and so we can replace $SpaceCost((\uparrow_a^w \bigcirc_a) \bowtie_{a,x} (\uparrow_x^y \bigcirc x))$ with $d_{out}^2 \cdot |V|$. For the operations cost we use $OpCost(A \bowtie B) = (SpaceCost(A) \cdot SpaceCost(B)) + OpCost(A) + OpCost(B)$, based on the complexity of a nested loop join (for simplicity). We also set $OpCost(\sigma A) = |A|$. The costs are then the following:

$$SpaceCost(\uparrow_z^u \circ \uparrow_w^z ((\uparrow_a^w \bigcirc_a) \bowtie_{a,x} (\uparrow_x^y \bigcirc x)))$$
$$= O(d_{out}^2 \cdot SpaceCost((\uparrow_a^w \bigcirc_a) \bowtie_{a,x} (\uparrow_x^y \bigcirc x)))(4.1.3)$$

$$OpCost(\sigma_K \circ \uparrow_z^u \circ \uparrow_w^z ((\uparrow_a^w \bigcirc_a) \bowtie_{a,x} (\uparrow_x^y \bigcirc x)))$$
$$= O((d_{out}^3 \cdot |V|) + (|E| \cdot (d_{out}^2 \cdot |V|)) + (|E| \cdot (|V| \cdot d_{out})) + ((d_{out}^2 \cdot |V|) \cdot (d_{out}^2 \cdot |V|))$$
$$+ (|E| \cdot (d_{out} \cdot |V|)) + (|E| \cdot (d_{out} \cdot |V|))) \qquad (4.1.4)$$

The above queries can be readily expanded to detect cycles of length greater than 4. In general, we could also consider squares with edges of any orientation by including both variants of the *expand* operator in our expression and filtering appropriately. Finally, we might be solely interested in cycles "centered" around a specific type of node, for instance nodes with a `name` attribute whose value is "`Alice`". In such a case, we can include an additional selection operation in the above trees, right after the bottom-most $\bigcirc_x$ operation.

4.2   Random Walk, Path, and Star-Path

Another frequently used graph query is that of performing an n-step random walk, starting from a specified node x. Such a query can be used to detect $s - t$ connectivity using very small amounts of space. To this end, we can introduce a new operator $RandRow(R)$ which selects a random row from a graph relation R. An example of a 2-step random walk, starting from a node with $id = ID$ is depicted by the left subfigure of Fig. 3.

Another interesting path operator is the *star-path*. In a star-path we wish to find a path between two nodes of interest and further want to expand the resulting graph relation to also include the neighbors of all nodes between the two terminals. We can make use of the random walk procedure to find n-length paths. For example, if we want paths of length $n = 2$, assuming the terminal nodes have ID_1, ID_2 respectively, there is only one inbetween node and our query can be seen in the right subfigure of Fig. 3. If successful (i.e., there exists a path of length 2 between the two terminals), it would add all the neighbors of the inbetween node under the attributes "a", for those connected to it via outgoing edges, and "b", for those connected via ingoing ones.

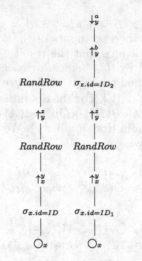

Fig. 3. A random 2-walk (left) and a star-path between ID_1 and ID_2 (right).

4.3 Grid Query

By appropriately combining multiple square queries we can create a new operator that detects grids in a graph. To do this, we require access to a function that,

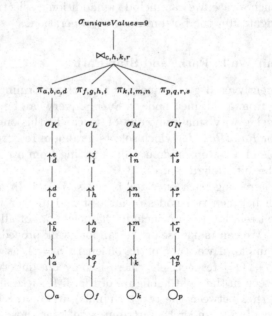

Fig. 4. Grid query, where $K \leftarrow a.id = e.id \wedge a.id \neq b.id \neq c.id \neq d.id$, $L \leftarrow f.id = j.id \wedge f.id \neq g.id \neq h.id \neq i.id$, $M \leftarrow k.id = o.id \wedge k.id \neq l.id \neq m.id \neq n.id$, $N \leftarrow p.id = t.id \wedge p.id \neq q.id \neq r.id \neq s.id$

given a graph relation R, adds a new attribute *uniqueValues* to it, containing
the number of unique values per row (see Fig. 4 for an example). Once again,
structuring our query appropriately is largely dependent on the specific edge
orientations we need to consider.

5 Experiments

5.1 Experimental Setup

We have implemented the aforementioned algebra of graph operators and relations in Python using Neo4j (community edition 3.1.2) as the graph database
and the Neo4j Python Driver `neo4j-driver` to facilitate communication with it.
The platform was setup on a 8GB, Ubuntu 16.04 VM. Data graphs were generated using the NetworkX Python library. The queries benchmarked consist of
a random 4-path and two square queries but modified to match a cycle with
vertex/edge pattern $x \leftarrow y \rightarrow z \leftarrow w \rightarrow x$ (see Fig. 5), so as to include both
ingoing and outgoing edges and be of more interest than the plain query shown
above. The actual time and space costs for queries presented in the benchmarks
were obtained by averaging over 10 random graph relations samples for each of
the queries executed. Queries that resulted in empty relations (i.e., happened
to start on nodes that no cycles where centered on) were discarded. The projected time and space costs were obtained by applying our cost model, using
node count, edge count, indegree/outdegree, etc., statistics provided by the NetworkX library (see Table 1). To compare the projected number of operations to
actual time costs in seconds, the projections were multiplied by constant factors.

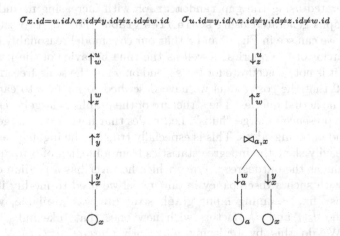

Fig. 5. The two square queries, thereafter referred to as SqrOne (left) and SqrTwo
(right).

Table 1. Statistics for graphs used in our benchmark, grouped by type of graph.

Type	Node count	Edge count	Average indegree/outdegree
Small Random	35	[242-1190]	[6.9143-34]
Dense Random	[1K-10K]	[5046 - 49639]	5.06
Sparse Random	[1K-10K]	[1998 - 19887]	2.04
Scale-Free	[1K-10K]	[2138 - 21439]	2.16

5.2　Results and Discussion

We now present the results of our benchmarks that compare the predictions of our cost model against actual costs obtained from Python implementations of each query over a Neo4j database.

We start with an idealized situation: Consider a small graph of fixed node count $n = 35$, where each node has a probability p to connect to any other node in the graph. By steadily increasing p we slowly approach a regular graph, a situation in which our model's predictions will theoretically perfectly match the actual results. For the two square queries mentioned before, we showcase the results in Fig. 6 (graphs generated using the `gnp_random_graph` function of the NetworkX Python Library, with increasing p). We note that the model accurately predicts size costs and also provides good results for the time costs.

Yet, a very small close-to-regular graph rarely reflects real life data. In the next benchmarks, we consider the projected and actual costs of two square queries on three sets of graphs: A pair of sparse (with average indegree/outdegree of around 2) and dense (average indegree/outdegree of around 5) random graphs, generated using the `gnp_random_graph` with increasing n, and a set of scale-free graphs generated using `scale_free_graph`[2]. For the two sets of random graphs, we can see in Figs. 7 and 8 that our cost model reasonably describes the size behavior of the queries as well as the time behavior of the first square query, while it is not as accurate for the second one. For the scale-free graphs, we note in Fig. 9 that the cost model we have described so far fails to capture the behavior of the actual queries. The structure of the graphs is largely responsible for this – the presence of large "hubs", i.e., nodes that have a very large number of ingoing and outgoing edges. This is especially true for the ingoing case, where such hubs heavily skew the indegree statistics from a median of 0 to an average of 2, the same as the outdegree average which however has a median of 1.

To deal with such a discrepancy in our model we need to modify it accordingly. An easy fix, assuming input graph structure was available, would be to replace the d_{out} and d_{in} factors with new ones that take hubs into consideration. We do this by replacing d_{out} with $OutGoing_{max}/d_{out}$ and d_{in} with $InGoingTrimmed_{max}/dTrimmed_{in}$, where $OutGoing_{max}$ is the maximum outdegree, $dTrimmed_{in}$ is the trimmed average of indegrees, and

---

[2] Using default arguments: `alpha=0.41`, `beta=0.54`, `gamma=0.05`, `delta_in=0.2`, `delta_out=0`.

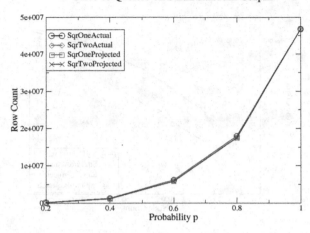

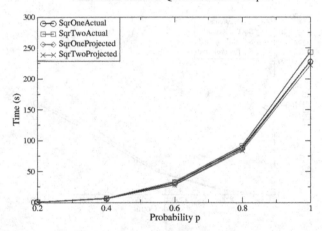

Fig. 6. Space and time costs (actual and projected) of the two square queries on small random graphs with increasing probability p of any two nodes connecting.

$InGoingTrimmed_{max}$ is the maximum indegree after *trimming*: We only need to "trim" the top 1% nodes (ordered by indegree) to obtain the predictions in Fig. 9. We can see that the modified costs more closely model the behavior of the actual queries and capture the difference between the two queries: The first one consumes more time and space than the second one, due to its use of more *expandOut* operations with high d_{out} factors as opposed to the second one which uses more *expandIn* operations.

Respective results for random four-walks follow the same general trends and can be found in the Appendix.

Sizes of Queries on Dense Random Graphs

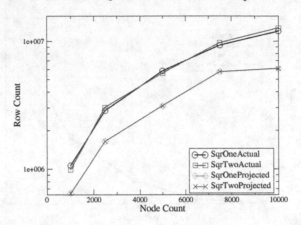

Execution Times of Queries on Dense Random Graphs

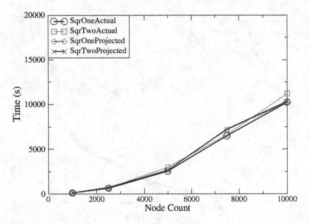

Fig. 7. Space and time costs (actual and projected) of the two square queries on dense random graphs with increasing N.

5.3 Including Label Information in the Cost Model

So far, we have discussed queries that only make use of global, non label-specific, statistics such as the average in/out-degree. However, it is also of interest to consider graphs that embellish edges with "label" or "type" data. To include such information in our cost models we must consider the cases where the labels are used in a query. Such a case, perhaps the most common one, is an expansion followed by a selection based on the label type:

$$query = \sigma_{xy.label=\text{``}labelType\text{''}} \circ \uparrow_x^y R \qquad (5.3.1)$$

In computing the costs of such a query, we can modify our cost model to include factors specific to the presence of labels. Note that the above query contains a

Sizes of Queries on Scale-Free Graphs

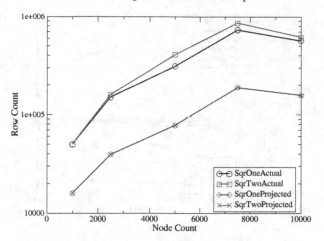

Execution Times of Queries on Sparse Random Graphs

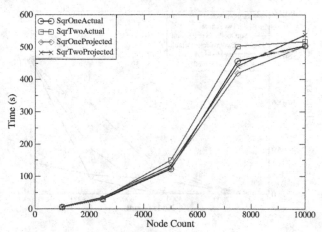

Fig. 8. Space and time costs (actual and projected) of the two square queries on sparse random graphs with increasing N.

selection operation, and as we discussed in Sect. 4.1, computing the space costs of a projection is very difficult unless we are supplied with relevant statistics which we can use to deduce costs. Fortunately, in this specific case, such information can be garnered from label-related statistics such as the in/out-going degree of a node per label type. As such, we can make use of this to modify our space cost as follows:

$$SpaceCost(query) = O(d_{out}^{labelType} |R|) \qquad (5.3.2)$$

where we use a superscript to denote the average degree as it pertains to a specific label type. As an example, suppose we take the sparse random graphs used in the previous experiments and assign to each edge a label $l \in \{label0, label1\}$ with

Resulting Sizes of Queries on Scale-Free Graphs

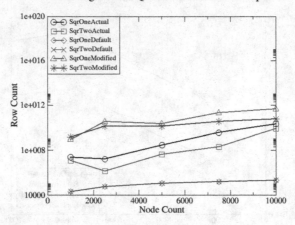

Execution Times of Queries on Scale-Free Graphs

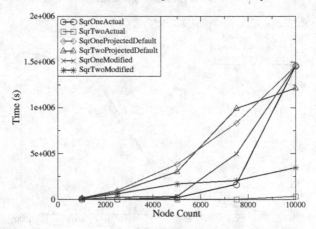

Fig. 9. Space and time costs (actual and projected via default and modified models) of the two square queries on scale-free graphs with increasing N.

a 0.3 probability of labeling it *label*0, and 0.7 of labeling it *label*1. We can then compute the relevant statistics of in/out-degrees per label, whose sums should agree with the "global" degrees d_{in}/d_{out} we've used before. Now consider the following query which makes use of a projection on the labeled edges:

$$query = \sigma_{xy.label="label0"} \circ \uparrow_x^y \circ \bigcirc_x \qquad (5.3.3)$$

Then Fig. 10 shows the relevant row-count benchmark of the actual query as well the projected costs from the default cost model and the cost model presented above. In general, it is quite more difficult to devise a cost model that includes label-related factors, as opposed to the label-agnostic one we have discussed earlier. One reason for that is the global behavior of the operators, in that they do not differentiate between labeled edges at all: *getNodes* fetches all nodes in a graph and *expandIn/expandOut* operate on all edges regardless of their label. Despite this difficulty, it is crucial to develop a cost model that includes such label-specific data, since much information about the characteristics of a graph can be found by examination of label-related metrics.

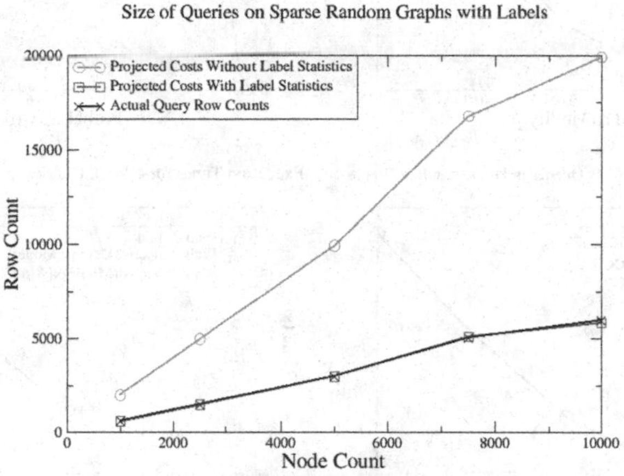

Size of Queries on Sparse Random Graphs with Labels

Fig. 10. Space costs (actual and projected) of query (5.3.3) using both the default cost model and the one modified to include label-related factors.

6   Conclusions

In this work we have presented an initial step towards creating algebraic cost models for various graph operators. Based on a simple model, we have demonstrated how various path operations can be decomposed and modeled utilizing the primitive operators, and how their costs can be computed. Our initial results show that, for a popular Graph DataBase, such modeling can quite accurately predict operator performance and cost. While implementation and internal computation models are important in devising model APIs, the structure and distribution of the graph itself can be of paramount importance. Thus, sampling, summarization techniques and sound knowledge in Network Science emerge as possibly critical factors towards the next steps in graph analytics modeling.

A   Appendix

A.1   Random 4-Walk Benchmarks

For the random 4-walks, we first note that $SpaceCost(RandRow(R)) = 1$, so final space costs collapse to 1. For the time costs, we present Fig. 11. We note that, in this case, both default and modified models fare well in their predictions – this can be attributed to the simplicity of the query and the fact that it uses only *expandOut* operations.

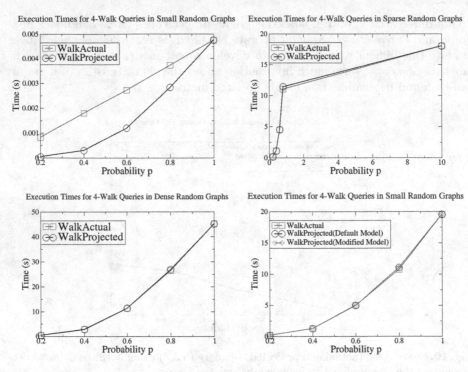

Fig. 11. Space and time costs (actual and projected via default and modified models) of 4-walk queries on graphs with various connectivity probabilities.

References

1. Apache Hama. https://hama.apache.org/
2. Apache Spark graphX. http://spark.apache.org/graphx/
3. Neo4j. https://neo4j.com/
4. Arenas, M., Gutierrez, C., Pérez, J.: Foundations of RDF databases. In: Tessaris, S., Franconi, E., Eiter, T., Gutierrez, C., Handschuh, S., Rousset, M.-C., Schmidt, R.A. (eds.) Reasoning Web 2009. LNCS, vol. 5689, pp. 158–204. Springer, Heidelberg (2009). https://doi.org/10.1007/978-3-642-03754-2_4
5. Cyganiak, R.: A relational algebra for SPARQL. Digital Media Systems Laboratory HP Laboratories Bristol. HPL-2005-170, vol. 35 (2005)
6. Doka, K., Papailiou, N., Giannakouris, V., Tsoumakos, D., Koziris, N.: Mix 'n' match multi-engine analytics. In: 2016 IEEE International Conference on Big Data, pp. 194–203 (2016)
7. Duggan, J., Elmore, A.J., Stonebraker, M., Balazinska, M., Howe, B., Kepner, J., Madden, S., Maier, D., Mattson, T., Zdonik, S.: The BigDAWG polystore system. In: ACM Sigmod Record (2015)
8. Frasincar, F., Houben, G.J., Vdovjak, R., Barna, P.: RAL: an algebra for querying RDF. World Wide Web **7**(1), 83–109 (2004)
9. Gonzalez, J., Low, Y., Gu, H., Bickson, D., Guestrin, C.: PowerGraph: distributed graph-parallel computation on natural graphs. In: Proceedings of the 10th USENIX Symposium on Operating Systems Design and Implementation (OSDI 12) (2012)

10. Hölsch, J., Grossniklaus, M.: An algebra and equivalences to transform graph patterns in neo4j. In: EDBT/ICDT 2016 Workshops: EDBT Workshop on Querying Graph Structured Data (GraphQ) (2016)
11. Kang, U., Tong, H., Sun, J., Lin, C.Y., Faloutsos, C.: GBASE: A scalable and general graph management system. In: Proceedings of the 17th ACM SIGKDD International Conference on Knowledge Discovery and Data Mining, KDD 2011 (2011)
12. LeFevre, J., Sankaranarayanan, J., Hacigumus, H., et al.: MISO: souping up big data query processing with a multistore system. In: Proceedings of the ACM SIGMOD International Conference on Management of Data (2014)
13. Papailiou, N., Tsoumakos, D., Karras, P., Koziris, N.: Graph-aware, workload-adaptive SPARQL query caching. In: Proceedings of the 2015 ACM SIGMOD International Conference on Management of Data, SIGMOD 2015 (2015)
14. Savnik, I., Nitta, K.: Algebra of RDF graphs for querying large-scale distributed triple-store. In: Buccafurri, F., Holzinger, A., Kieseberg, P., Tjoa, A.M., Weippl, E. (eds.) CD-ARES 2016. LNCS, vol. 9817, pp. 3–18. Springer, Cham (2016). https://doi.org/10.1007/978-3-319-45507-5_1
15. Schmidt, M., Meier, M., Lausen, G.: Foundations of SPARQL query optimization. In: Proceedings of the 13th International Conference on Database Theory, pp. 4–33. ACM (2010)
16. Yan, D., Bu, Y., Tian, Y., Deshpande, A., Cheng, J.: Big graph analytics systems. In: Proceedings of the 2016 International Conference on Management of Data, SIGMOD 2016 (2016)

Computing Probabilistic Queries in the Presence of Uncertainty via Probabilistic Automata

Theodore Andronikos, Alexander Singh, Konstantinos Giannakis[✉],
and Spyros Sioutas

Department of Informatics, Ionian University, Corfu, Greece
{andronikos,p13sing,kgiann,sioutas}@ionio.gr

Abstract. The emergence of uncertainty as an inherent aspect of RDF
and linked data has spurred a number of works of both theoretical and
practical interest These works aim to incorporate such information in a
meaningful way in the computation of queries. In this paper, we propose
a framework of query evaluation in the presence of uncertainty, based
on probabilistic automata, which are simple yet efficient computational
models. We showcase this method on relevant examples, where we show
how to construct and exploit the convenient properties of such automata
to evaluate RDF queries with adjustable cutoff. Finally, we present some
directions for further investigation on this particular line of research,
taking into account possible generalizations of this work.

Keywords: Probabilistic linked data · SPARQL queries
Probabilistic queries

1 Introduction and Motivation

Linked Data is an umbrella term that describes the methodology of generating,
sharing, and connecting data and information on the WWW. In the direction
of making data easy to be discovered and manipulated by the community and
organizations [16], the need for new, enhanced languages and frameworks is
highlighted. The most widely used framework is the RDF and for the querying
part the SPARQL language [1]. The network of these interlinked data sources
forms the so-called Web of Linked Data. In principle, research on Linked Data
concerns procedures on the producing, storing, and querying of usually big sets
of data.

The importance of linked data for representing knowledge cannot be over-
stated. However, at least in certain domains, the facts expressed by linked data
are not 100% certain. These facts can be characterized by an extra parameter
which expresses the confidence in their correctness. Typically this parameter
is a real number taking values in the interval $[0, 1]$ and it can be thought of
as the *probability* that the triple in question is correct. Equivalently, it can be
considered as the *weight* or *degree of certainty*.

© Springer International Publishing AG, part of Springer Nature 2018
D. Alistarh et al. (Eds.): ALGOCLOUD 2017, LNCS 10739, pp. 106–120, 2018.
https://doi.org/10.1007/978-3-319-74875-7_7

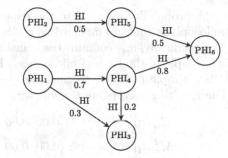

```
SELECT ?phil
WHERE {
    Philosopher1 HasInfluenced+ ?phil
}
```

Fig. 1. On the left, a SPARQL query that lists the philosophers influenced by "Philosopher1." On the right, an RDF graph depicting the influences among a group of philosophers. Each node represents a Philosopher. The predicates labeling every edge express the property HasInfluenced, whereas the values express the uncertainty weight.

Example 1. To develop the intuition for this situation, let us consider the RDF graph depicted in Fig. 1. The underlying set of triples $\mathcal{D}$ is about philosophers, or thinkers in general, and the possible influences among them. The label PHI_i is understood as an abbreviation for "Philosopher i." Moreover, we take for granted the existence of a probability measure $\mathbf{P}$ that assigns to each $l \in \mathcal{D}$ a real number $\mathbf{P}(l)$ in the interval $[0, 1]$.

For instance, $\mathcal{D}$ could contain the triple (Kant, HasInfluenced, Schopenhauer) to which the probability value 0.7 may be assigned. We further assume that the "influence" property, expressed by the predicate "HasInfluenced," is transitive in nature, i.e., if (PhilosopherA, HasInfluenced, PhilosopherB) and (PhilosopherB, HasInfluenced, PhilosopherC), then, at least indirectly, PhilosopherA has influenced PhilosopherC. The query in Fig. 1 will list the philosophers in $\mathcal{D}$ that have been influenced by "Philosopher 1." In this query we have used the syntax of SPARQL 1.1 [1] that allows us to express path properties; here the symbol + means one or more transitions labeled with the predicate "HasInfluenced."

$$M_{HI} = \begin{bmatrix} 0.0 & 0.0 & 0.0 & 0.0 & 0.0 & 0.0 \\ 0.0 & 0.0 & 0.0 & 0.0 & 0.0 & 0.0 \\ 0.3 & 0.0 & 0.0 & 0.2 & 0.0 & 0.0 \\ 0.7 & 0.0 & 0.0 & 0.0 & 0.0 & 0.0 \\ 0.0 & 0.5 & 0.0 & 0.0 & 0.0 & 0.0 \\ 0.0 & 0.0 & 0.0 & 0.8 & 0.5 & 0.0 \end{bmatrix} \quad \text{and} \quad \mathbf{q_1} = \begin{bmatrix} 1 \\ 0 \\ 0 \\ 0 \\ 0 \\ 0 \end{bmatrix}. \tag{1}$$

The matrix M_{HI} encodes the information shown in Fig. 1 and the vector $\mathbf{q_1}$ represents node 1, corresponding to "Philosopher 1." Matrix and vector representations are particularly convenient, since they facilitate in a uniform and succinct way the operations we perform upon the probabilistic linked data. Technically, M_{HI} corresponds to the predicate HasInfluenced. The probabilities assigned to the triples are elements of the matrix M_{HI}. We note that the element m_{ij} of matrix M_{HI} contains the probability that node j is connected to node i: m_{ij}

is the probability assigned to the triple $(j, \text{HasInfluenced}, i)$, where j and i are philosophers that are modeled as nodes.

Specifically, m_{i1} contains the transition probability from 1 to node i. If there is no triple of the form $(\text{Philosopher1, HasInfluenced}, i)$ then m_{i1} is equal to 0. It is a simple task now to calculate the products $M_{HI}\mathbf{q}_1, M_{HI}^2\mathbf{q}_1$ and $M_{HI}^3\mathbf{q}_1$. The resulting vectors are the following:

$$M_{HI}\mathbf{q}_1 = \begin{bmatrix} 0.00 & 0.00 & 0.30 & 0.70 & 0.00 & 0.00 \end{bmatrix}^T, \tag{2}$$

$$M_{HI}^2\mathbf{q}_1 = \begin{bmatrix} 0.00 & 0.00 & 0.14 & 0.00 & 0.00 & 0.56 \end{bmatrix}^T, \quad \text{and} \tag{3}$$

$$M_{HI}^3\mathbf{q}_1 = \begin{bmatrix} 0.00 & 0.00 & 0.00 & 0.00 & 0.00 & 0.00 \end{bmatrix}^T. \tag{4}$$

The interpretation of these results is quite straightforward. The i-th element of $M_{HI}\mathbf{q}_1$ is the probability that "Philosopher 1" influenced "Philosopher i." This information is contained in the dataset. In general the i-th element of $M_I^n\mathbf{q}_1$ is the probability that "Philosopher 1" influenced "Philosopher i" via $n-1$ philosophers $k_1, \ldots, k_{n-1}$, i.e., the probability of the path $1, k_1, \ldots, k_{n-1}, i$. There is no point going further after encountering a vector with all entries 0, as can immediately be seen from the graph itself. In case of acyclic graphs, such as this one, the total influence is found by adding $M_{HI}\mathbf{q}_1 + M_{HI}^2\mathbf{q}_1 = \begin{bmatrix} 0.00 & 0.00 & 0.44 & 0.70 & 0.00 & 0.56 \end{bmatrix}^T$. Thus, we see that "Philosopher 1" has influenced "Philosopher 3", "Philosopher 4" and "Philosopher 6." We may, however, consider a situation where in order to accept an inferred fact, its corresponding probability has to be above a certain threshold; e.g., if this threshold is 0.5, then we *cannot* conclude that "Philosopher 1" has influenced "Philosopher 3." □

SPARQL is capable of querying diverse data sources and derived results are defined through basic graph pattern matching [10]. It has recently been used for querying structures consisting of uncertain Linked Data. Many actual applications and domains have an inherently probabilistic nature. Large volumes of data are both generated and described under uncertainty. Biology and biological data form an eminent domain where one meets stochasticity. For example, probabilistic links between biological concepts can be derived from several prediction techniques and probabilistic links are mutually independent [12].

The above observations highlight the need for mechanisms to address the management and control of such resources. In this direction, probabilistic or uncertain data in RDF form have to be processed using analogous tools. It is obvious that in a classical, traditional database system, the result of a query is distinctive, whereas in a probabilistic database, the underlying system should make calculations that correspond to each answer depending on probability values that represent the uncertainty levels. The work of [12] was one of the first to discuss the querying mechanism of probabilistic RDF databases, using a proposed framework for the support of SPARQL. The problem of querying uncertain linked data has also been investigated in [11,15]. Uncertain or probabilistic graphs could also be used for the representation and analysis of noisy linked data which occur in many real-world scenarios [13], such as the automated generation of linked data parts (like the proposed methodology of [22]).

A regular path query (RPQ) asks for pairs of nodes in the graph, connected by a path, which can be expressed in terms of a regular expression. These expressions form a subset of the class of regular expressions. The execution of the RPQ involves a sequence of edges of the graph that are followed in order to reach the desired result. The question now becomes how can one compute SPARQL queries encompassing uncertainty. The methodology we propose in this work is to use probabilistic automata. For certain classes of queries one can design a corresponding probabilistic automaton.

By adjusting the threshold of the automaton, one can compute those query answers for which the certainty is above this threshold. This approach is more suitable compared with mainstream methods that include the *a posteriori* calculation of the derived probabilistic result. The use of an "inline" approach for producing the result could enhance the performance in cases where multiple redundant path would have been processed (i.e., paths with zero weights). Another advantage of the proposed method concerns the cases where the weights are easily (or constantly) altered in the runtime.

Contribution. Summarizing the contributions of this paper, we first note that our effort shows how to assign a probabilistic automaton corresponding to a particular query on RDF-like data structures. Our approach is motivated by the need to develop mechanisms for querying data in the presence of uncertainty, where there is a probabilistic aspect present in each triple. This is, to the best of our knowledge, the first time that probabilistic automata have been proposed for this particular task. The wealth of well-studied results on these abstract formal tools, as well as the practical efficiency outlined in the previous paragraph, constitute the main advantages of our method. Furthermore, this approach can be readily extended to queries on more complex variants in a potentially infinite web of linked data.

The paper is structured as follows: the most relevant works are discussed in Sect. 2, whereas Sect. 3 contains the necessary definitions and notation used throughout this paper. The main contribution of this work is presented and analyzed in Sect. 4. Finally, conclusions and directions for future work are reported in Sect. 5.

2   Related Work

Recently, Zhang et al. in [24] expanded navigational path queries using context-free notions, rather than regular expressions. Their primary motivation was the restricted expressiveness of the latter, for example in the case of same generation-queries. They introduce the so-called cfSPARQL, which acts as an extension of the standard SPARQL equipped with context-free grammars. Using cfSPARQL, one can pose more sophisticated queries. The computational complexity does not scale up, and the overall query evaluation is efficient for most cases.

Sistla et al. in [21] established an important connection between automata and database queries. In particular, they described a novel way for specifying queries over sequence databases based on automata. They focus specifically on similarity issues, such as similarity-based retrieval from databases for various distance metrics. These systems are capable of answering nearest neighbor or range queries.

Optimization is needed when one considers queries with regular expressions and several works have been proposed. E.g., Fernandez and Suciu in [7] introduce two graph algorithms for specifying partial knowledge about the data's structure, whereas Barceló et al. examine graph pattern queries using notions from automata theory [4]. An algorithm based on automata for answering specific RDF queries was proposed by Wang et al. in [23]. These queries are regular path queries that take into account information about the provenance. Their test results revealed that their method is able to effectively and efficiently answer provenance-aware regular path queries on large real-world RDF graphs.

Similarly to our approach, Hua and Pei discussed probabilistic path queries [11]. Their motivation and initial target is similar to ours, but they do not use an automata-like tool. Rather, they adopted a method based on dynamic programming, using well-known techniques combined with efficient heuristics. One of the first works on probabilistic RDF databases and the associated querying process was carried out by Huang and Liu in [12]. Besides providing the motivation for such an approach, they describe a related framework that supports SPARQL queries for probabilistic databases. One of their main results was an approximation algorithm for computing path expressions in an efficient manner, demonstrated and evaluated through experiments.

RDF graphs with uncertainty were discussed by Lian et al. [15]. They added probability values to standard RDF data, transforming them into probabilistic RDF graphs. They addressed particular problems about retrieving subgraphs from inconsistent probabilistic RDF graphs that are isomorphic to a given query graph with high quality scores. They presented two pruning methods: adaptive label pruning and quality score pruning. Another work that discussed uncertainty in Linked Data, was the one by Reynolds [19]. His theoretical approach emphasized the need for considering uncertain data and proposed solutions to corresponding issues.

Another important work on querying probabilistic databases was the one by Dalvi and Suciu [5], in particular on the evaluating aspect of the queries, using an optimization algorithm. The proposed system is capable of supporting complex SQL queries, using probabilistic query semantics and the results are ranked. A novel query language called pSPARQL was introduced by Fang and Zhang [6]. This language is designed for querying probabilistic RDF and it is built on SPARQL.

Akbarinia et al. also consider probabilistic databases in their study of SUM queries (i.e., queries that return the sum of given values), where they try to tackle the known problem of small aggregate values as a result of probability calculations [2]. A similar problem is addressed by Krompass et al. using a different approach [14]. Recently, Schoenfisch in [20] discussed the correlation of ontologies with probabilistic character and he described the aspect of posing SPARQL queries on such data structures.

One of the first attempts to correlate automata variants to queries on RDF-like data was undergone in [8]. In this work, however, the main scope was the infinite character of the dynamically increasing size of Linked Data sources which need appropriate handling. Similarly, Giannakis et al. continued this thought and extended their rationale achieving the association of particular queries on transitive labels with ω-regular languages, which is the extension of standard regular languages in the case of infinite inputs [9]. One could see the work described here as an extension of the above, taking this time into account the uncertainty that has been introduced to our data stores.

3   Definitions and Notation

The standard references for probabilistic automata are the landmark paper of Rabin [18] and the book of Paz [17]. We begin by defining when a vector of real numbers is stochastic or substochastic. A vector $\mathbf{v} = \begin{bmatrix} a_1 & \cdots & a_n \end{bmatrix}^T$ is *stochastic*, if $a_i \geq 0$, $1 \leq i \leq n$, and $\sum_{i=1}^{n} a_i = 1$; $\mathbf{v} = \begin{bmatrix} a_1 & \cdots & a_n \end{bmatrix}^T$ is *substochastic*, if $a_i \geq 0$, $1 \leq i \leq n$, and $\sum_{i=1}^{n} a_i \leq 1$. Although $\mathbf{v}$ is a column vector, it is more economical in terms of space to express it as the transpose of a row vector. These notions can be extended to square matrices. An $n \times n$ matrix M is *stochastic* (*substochastic*) if all its columns are stochastic (respectively substochastic).

Definition 1. *A probabilistic finite automaton (PFA) is a tuple $(Q, \Sigma, \mathcal{M}, q_0, F)$ where:*

(1) Q is a finite set of states.
(2) Σ is the finite input alphabet.
(3) $\mathcal{M}$ is a finite set of $|\Sigma|$ transition matrices. Each transition matrix is a square $|Q| \times |Q|$ substochastic matrix. There is 1-1 correspondence between the letters of Σ and the transition matrices of $\mathcal{M}$; for each $a \in \Sigma$, the corresponding $M_a \in \mathcal{M}$ contains the probabilities that the automaton will move from one state to another upon reading letter a.
(4) $q_0 \in Q$ is the initial state.
(5) $F \subseteq Q$ is the set of accepting states.

The set of all words over the alphabet Σ is denoted by $\Sigma^\star$. The following example will clarify the operation of a probabilistic automaton and elaborate the effect of the transition matrices on the acceptance or rejection of a given word.

Example 2. Consider the automaton A shown in Fig. 2. A has four states: q_1, q_2, q_3 and q_4, q_1 is the initial state (indicated by the small arrow pointing to q_1) and q_4 is the unique accepting state (indicated by the two concentric cycles). The transitions between the states are labeled with input letters followed by the corresponding probability. For instance, there is a transition from q_1 to q_2 labeled with $a : 0.5$, implying that whenever the automaton is at state q_1 and scans the letter a, there is a 0.5 probability to enter state q_2. Similarly, there is a second transition from q_1 to q_2 labeled with $b : 1.0$. In this case, the

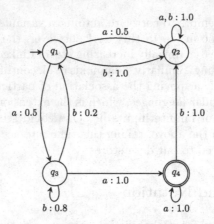

Fig. 2. An example of a probabilistic automaton.

automaton will surely move from q_1 to q_2, i.e., with probability 1.0, upon reading b. From the Fig. 2 we also infer that the input alphabet Σ is $\{a, b\}$.

The probabilistic behavior of A is encoded in the two stochastic matrices

$$
M_a = \begin{bmatrix} 0.0 & 0.0 & 0.0 & 0.0 \\ 0.5 & 1.0 & 0.0 & 0.0 \\ 0.5 & 0.0 & 0.0 & 0.0 \\ 0.0 & 0.0 & 1.0 & 1.0 \end{bmatrix} \text{ and } M_b = \begin{bmatrix} 0.0 & 0.0 & 0.2 & 0.0 \\ 1.0 & 1.0 & 0.0 & 0.0 \\ 0.0 & 0.0 & 0.8 & 0.0 \\ 0.0 & 0.0 & 0.0 & 1.0 \end{bmatrix}. \tag{5}
$$

We point out that the elements m_{ij}^a and m_{ij}^b of the matrices M_a and M_b respectively, contain the probabilities of the transitions from j to i upon reading letters a and b respectively. The states are represented by unit vectors: $\mathbf{q}_1$ corresponds to $\begin{bmatrix} 1 & 0 & 0 & 0 \end{bmatrix}^T$, $\mathbf{q}_2$ to $\begin{bmatrix} 0 & 1 & 0 & 0 \end{bmatrix}^T$, $\mathbf{q}_3$ to $\begin{bmatrix} 0 & 0 & 1 & 0 \end{bmatrix}^T$ and $\mathbf{q}_4$ to $\begin{bmatrix} 0 & 0 & 0 & 1 \end{bmatrix}^T$. To each input word $w = a_1 \ldots a_n \in \Sigma^*$ we associate the matrix product $M_w = M_{a_n} \ldots M_{a_1}$. Note that the order of the transition matrices in the product M_w is the *reverse* of the order of the letters in w.

Assuming that A is initially at state q_1, then after reading the input word w, the exact probabilistic distribution of the states A may end up is given by the stochastic vector

$$
\mathbf{p} = \begin{bmatrix} p_1 & p_2 & p_3 & p_4 \end{bmatrix}^T = M_w \mathbf{q}_1. \tag{6}
$$

Actually, we are only interested in the probability p_4 that A ends up in the accepting state q_4 because a word w is *accepted* only if the automaton upon reading w ends up in one of its accepting states. So, we may write

$$
p_4 = \begin{bmatrix} 0 & 0 & 0 & 1 \end{bmatrix} M_w \mathbf{q}_1. \tag{7}
$$

Suppose that A is fed a word beginning with b, like bb, ba. Using Eq. (7) we compute that the probability that A will end up in state q_4 is 0. If we use Eq. (6) we see that in both cases A will end up in state q_2 with probability 1.0.

Any word beginning with b will surely end up in q_2. In view of the fact that q_2 is not an accepting state, we may conclude that such words are not accepted by A. On the other hand, if A reads a word consisting entirely of a, like aa or $aaaa$, it will end up in the accepting state q_4 with probability 0.5. These words are accepted by A, provided that the value 0.5 is considered as an adequate degree of certainty. If, however, the confidence requirements are higher, let's say 0.75, then these words are not accepted either.   □

Given a probabilistic automaton A, let $P_A(w)$ denote the probability that A ends up in an accepting state upon scanning input word w. The language accepted by A is defined in terms of real parameter called cut-point (or threshold) and the languages recognized by probabilistic automata are called stochastic.

Definition 2. *Let A be a probabilistic finite automaton and let λ be a real number such that $0 \leq \lambda < 1$. The language accepted by A with cut-point λ, denoted by $L_A(\lambda)$, is defined as follows:*

$$L_A(\lambda) = \{w \in \Sigma^* | P_A(w) > \lambda\}. \tag{8}$$

A word w is accepted (or recognized) by A with cut-point λ, if $P_A(w) > \lambda$.

An RDF (linked data) *triple* has the standard syntax (`subject, predicate, object`) and is usually abbreviated as (s, p, o). The intended semantics of such triples is that the subject has the relationship described by the predicate with the object. Geometrically, a triple can be pictured as two adjacent vertices of a graph labeled with s and o respectively, together with a transition (directed edge) from s to o labeled with p. In the presence of uncertainty it is not 100% certain that the asserted relationship between the subject and the holds. Instead we assume that each triple ia assigned a real number in the interval $[0, 1]$ indicating the confidence of its assertion. This situation is described in the next definition.

Definition 3. *Let $\mathcal{U}$, $\mathcal{B}$, $\mathcal{V}$ and $\mathcal{L}$ denote pairwise disjoint sets of URIs, blank nodes, variables and literals, respectively. An RDF triple is a triple (s, p, o), where $s \in (\mathcal{U} \cup \mathcal{B})$, $p \in \mathcal{U}$ and $o \in (\mathcal{U} \cup \mathcal{B} \cup \mathcal{L})$.*
A probabilistic dataset, denoted $\mathcal{D}_\mathbf{P}$ is a finite set $\mathcal{D}$ of RDF triples endowed with a function $\mathbf{P} : \mathcal{D} \rightarrow [0, 1]$ that assigns to each $l \in \mathcal{D}$ the real number $\mathbf{P}(l) \in [0, 1]$.

In this work we view the numbers assigned to triples by $\mathbf{P}$ as probabilities, but they may equally well be considered as the *weights* or the *degrees of certainty* of each triple in $\mathcal{D}$.

4   Using Probabilistic Automata to Answer Queries

In previous sections we have presented the necessary background and definitions. Here we proceed with our main contribution: how probabilistic automata can facilitate the evaluation of SPARQL queries over probabilistic linked data. The connection between SPARQL queries and automata has been established in previous works [8,9]. In this work we advocate a novel approach regarding the use of probabilistic automata as appropriate tools for computing SPARQL queries in the presence of uncertainty.

Automata are simple yet powerful models of computation that have been widely studied since their initial description. The relevant literature is full of important results and, in many cases, efficient algorithms. Data querying requires effective and easy to implement algorithms, something that seems to bode well with automata-theoretic methods.

In dealing with RDF graphs it will be convenient, if not necessary, to consider nodes with no outgoing transitions, or nodes for which the probabilities of the outgoing transitions labeled with a specific predicate do not sum up to 1, but to some positive real value less than 1. It is precisely for these pragmatic reasons that in Definition 1 the transition matrices are required to be *substochastic*. In our context, if a state has no outgoing transitions for a given letter a, then the corresponding column in the transition matrix M_a will contain only zeroes.

This means that the underlying graphs are quite sparse. In fact, this is a crucial parameter, since it is possible to reduce the space complexity using well-known techniques from numerical analysis. Also, the fact that the paths are weighted with probabilities means that the actual querying part concerns a small subgraph of the overall system. In many cases products of probabilistic values converge rapidly to 0. In a real-world application where the acceptable threshold is much greater than 0, this can be used to speed up the computation by quickly abandoning paths that fall below the threshold. In marginal cases where the threshold is close to 0, we can adopt the use of simple NFA instead of PFA.

Here, we state two assumptions that underlie our approach. The first assumption is that all triples l in our dataset $\mathcal{D}$ are assigned probabilities; formally, there is a mapping $\mathbf{P} : \mathcal{D} \rightarrow [0,1]$ that assigns to each $l \in \mathcal{D}$ the real number $\mathbf{P}(l) \in [0,1]$. The second assumption is that all triples in our universe of discourse are probabilistically independent. This assumption will enable us to infer new probabilities from the existing probabilities attributed to the linked data.

E.g., suppose that the probabilities assigned to (TownA, LocatedIn, RegionB) and (RegionB, LocatedIn, RegionC) are 0.9 and 0.8 respectively. By assuming that the aforementioned triples are independent, we may then infer that (TownA, LocatedIn, RegionC) with probability $0.9 \times 0.8 = 0.72$. The independence assumption corresponds to many real-world graphs, mainly inspired from the biological domain, although there are also cases where this observation obviously does not hold.

4.1   Constructing Automata from Queries

We begin this part with an example that will illustrate the details of our proposed technique and then we shall give an abstract and general description of the method.

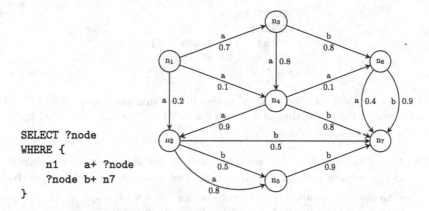

```
SELECT ?node
WHERE {
    n1      a+ ?node
    ?node b+ n7
}
```

Fig. 3. On the left, a SPARQL query that lists the nodes that are accessible from "n1" via an a path and from which "n7" is accessible via a b path. On the right, an RDF graph containing two transitive predicates labeled a and b. The probability of each transition is indicated below the predicate labeling the transition.

Example 3. Consider the RDF graph and the SPARQL query of Fig. 3. We assume that labels a and b stand for transitive predicates. The choice of transitive predicates and the utilization of the enhanced navigational properties of SPARQL 1.1 [1] will allow us to better illustrate the capabilities of probabilistic automata. We will use the term a path (b path) for a path consisting entirely of directed edges labeled with the same predicate a (b). This query will return *all* nodes, designated by the variable "?node," that satisfy the following 2 requirements with probability at least p.

(**R1**) There is an a path of length at least one from "n1" to "?node", and
(**R2**) there is a b path of length at least one from "?node" to "n7".

We may assume that SPARQL 1.1 is extended so as to allow us to specify that the answer to the query must hold with probability at least p. Unfortunately, the current version of SPARQL 1.1 lacks this functionality. In a world of absolute certainty, the evaluation of the above query should produce all the nodes for which the above two conditions hold. However, in a probabilistic setting, where every triple has an associated degree of certainty, the query should return only those nodes for which the probability that they meet both requirements simultaneously is *at least p*.

Typically p will be a real number in the interval $[0, 1]$. The higher p is, the fewer the nodes, and in the limiting case where $p = 0$ the situation is exactly as in the non-probabilistic case. The query in Fig. 3 is conjunctive, i.e., it is the conjunction of two queries: the first computes the nodes satisfying (**R1**) and the second those satisfying (**R2**). A node "?node" will be included in the answer if it satisfies both (**R1**) and (**R2**) with probabilities p_1 and p_2, respectively, and $p_1 p_2 > p$. One could also consider an alternative formulation of the query of Fig. 4. In this form, an answer node should fulfill criteria (**R1**) and (**R2**) with probability at least p_1 and p_2, respectively.

```
SELECT ?node
WHERE {
        n1      a+ ?node [p1]
        ?node b+ n7 [p2]
}
```

Fig. 4. This query will list "?node" if the probability of an a path from "n1" to "?node" is at least p_1 and the probability of a b path from "?node" to "n7" is at least p_2.

The transition matrices M_a and M_b contain all the information shown in Fig. 3 and the column vectors q_1 and q_7 represent nodes n_1 and n_7, respectively.

$$M_a = \begin{bmatrix} 0.0 & 0.0 & 0.0 & 0.0 & 0.0 & 0.0 & 0.0 \\ 0.2 & 0.0 & 0.0 & 0.9 & 0.0 & 0.0 & 0.0 \\ 0.7 & 0.0 & 0.0 & 0.0 & 0.0 & 0.0 & 0.0 \\ 0.1 & 0.0 & 0.8 & 0.0 & 0.0 & 0.0 & 0.0 \\ 0.0 & 0.8 & 0.0 & 0.0 & 0.0 & 0.0 & 0.0 \\ 0.0 & 0.0 & 0.0 & 0.1 & 0.0 & 0.0 & 0.0 \\ 0.0 & 0.0 & 0.0 & 0.0 & 0.0 & 0.4 & 0.0 \end{bmatrix} \quad M_b = \begin{bmatrix} 0.0 & 0.0 & 0.0 & 0.0 & 0.0 & 0.0 & 0.0 \\ 0.0 & 0.0 & 0.0 & 0.0 & 0.0 & 0.0 & 0.0 \\ 0.0 & 0.0 & 0.0 & 0.0 & 0.0 & 0.0 & 0.0 \\ 0.0 & 0.0 & 0.0 & 0.0 & 0.0 & 0.0 & 0.0 \\ 0.0 & 0.5 & 0.0 & 0.0 & 0.0 & 0.0 & 0.0 \\ 0.0 & 0.0 & 0.8 & 0.0 & 0.0 & 0.0 & 0.0 \\ 0.0 & 0.5 & 0.0 & 0.8 & 0.9 & 0.9 & 0.0 \end{bmatrix} \quad q_1 = \begin{bmatrix} 1 \\ 0 \\ 0 \\ 0 \\ 0 \\ 0 \\ 0 \end{bmatrix} \quad q_7 = \begin{bmatrix} 0 \\ 0 \\ 0 \\ 0 \\ 0 \\ 0 \\ 1 \end{bmatrix}$$

$$(9)$$

Suppose that we want to compute the query in Fig. 3 with probability 0.5. Using the theory of probabilistic automata that we explained in Sect. 3, we calculate the products $M_a^n q_1$, $1 \leq n \leq 5$, and $q_7^T M_b^m$, $1 \leq m \leq 3$. The i-th element of $M_a^n q_1$ gives the probability that there is an a path of length n from "n1" to node n_i. The j-th element of $q_7^T M_b^m$ gives the probability that there is a b path of length m from node n_j to "n7."

$$M_a q_1 = \begin{bmatrix} 0.000 & 0.200 & 0.700 & 0.100 & 0.000 & 0.000 & 0.000 \end{bmatrix}^T, \tag{10}$$

$$M_a^2 q_1 = \begin{bmatrix} 0.000 & 0.090 & 0.000 & 0.560 & 0.160 & 0.001 & 0.000 \end{bmatrix}^T, \tag{11}$$

$$M_a^3 q_1 = \begin{bmatrix} 0.000 & 0.504 & 0.000 & 0.000 & 0.072 & 0.056 & 0.004 \end{bmatrix}^T, \tag{12}$$

$$M_a^4 q_1 = \begin{bmatrix} 0.000 & 0.000 & 0.000 & 0.000 & 0.403 & 0.000 & 0.022 \end{bmatrix}^T, \tag{13}$$

$$M_a^5 q_1 = \begin{bmatrix} 0.000 & 0.000 & 0.000 & 0.000 & 0.000 & 0.000 & 0.000 \end{bmatrix}^T, \tag{14}$$

$$q_7^T M_b = \begin{bmatrix} 0.000 & 0.500 & 0.000 & 0.800 & 0.900 & 0.900 & 0.000 \end{bmatrix}^T, \tag{15}$$

$$q_7^T M_b^2 = \begin{bmatrix} 0.000 & 0.450 & 0.720 & 0.000 & 0.000 & 0.000 & 0.000 \end{bmatrix}^T, \text{ and} \tag{16}$$

$$q_7^T M_b^3 = \begin{bmatrix} 0.000 & 0.000 & 0.000 & 0.000 & 0.000 & 0.000 & 0.000 \end{bmatrix}^T. \tag{17}$$

This process stops when the probability vectors become zero. By adding them, we obtain $\mathbf{p}_a$ and $\mathbf{p}_b$ whose i-th element gives the probability of an a path (of arbitrary length) from "n1" to n_i and the probability of a b path from n_i to "n7," respectively. Thus, the probability that there is an a path from "n1" to n_i and in the same time from n_i to "n7" is given by the product of the corresponding elements.

$$\mathbf{p}_a = \begin{bmatrix} 0.000 & 0.794 & 0.700 & 0.660 & 0.635 & 0.066 & 0.026 \end{bmatrix}^T, \tag{18}$$

$$\mathbf{p}_b = \begin{bmatrix} 0.000 & 0.950 & 0.720 & 0.800 & 0.900 & 0.900 & 0.000 \end{bmatrix}, \tag{19}$$

$$\mathbf{p} = \begin{bmatrix} \mathbf{p}_{a,1}\mathbf{p}_{b,1} & \mathbf{p}_{a,2}\mathbf{p}_{b,2} & \mathbf{p}_{a,3}\mathbf{p}_{b,3} & \mathbf{p}_{a,4}\mathbf{p}_{b,4} & \mathbf{p}_{a,5}\mathbf{p}_{b,5} & \mathbf{p}_{a,6}\mathbf{p}_{b,6} & \mathbf{p}_{a,7}\mathbf{p}_{b,7} \end{bmatrix} \tag{20}$$

$$= \begin{bmatrix} 0.000 & 0.754 & 0.504 & 0.528 & 0.571 & 0.059 & 0.000 \end{bmatrix}^T. \tag{21}$$

The answer to the query in Fig. 3 is contained in $\mathbf{p}$. In view of the actual numerical values given in (21), the query (for threshold 0.5) will output nodes n_2, n_3, n_4 and n_5. If the threshold is increased to a higher value, e.g., 0.75, then the query will return only n_2. Vectors $\mathbf{p}_a$ and $\mathbf{p}_b$ have all the information required to handle the query in Fig. 4. If for example $p_1 = 0.7$ and $p_2 = 0.7$ then this query will produce nodes n_2 and n_3.

Let us a point out that certain levels of optimization can be carried out to speed up the computation process. For instance an upper bound on the length of the path may be imposed, so as to reduce the number of matrix-vector multiplications. Furthermore, any node for which the total path probability falls under the threshold may be safely ignored. The nature of probabilistic computations guarantees that the probability will never increase in a subsequent step of the computation.

For instance, if the probability threshold is 0.7 and we have first computed vector $\mathbf{p}_a$, then we can be certain that only n_2 and n_3 may satisfy the query; the probability for all other nodes will be below 0.7, as they are will be multiplied by a number at most equal to one. Hence, the other nodes need not be considered anymore.   □

4.2 The General Method

We shall now outline in clear and succinct steps the proposed method for constructing a probabilistic automaton that computes a SPARQL conjunctive query over an RDF graph (see Fig. 5). We use the following terminology: a node "node1" is an *initial* node if it appears syntactically as `node1 a+ ?node`, a node "node2" is a *final* node if it appears syntactically as `?node a+ node2`. We note that our method of checking the probabilities against a threshold also guards against cases where the structure of the graph itself, for example in the presence of loops, causes the query probabilities to drop too low, obscuring the results.

1: ◁ INPUT: An RDF graph G and a SPARQL query ▷
2: ◁ OUTPUT: A probabilistic automaton A ▷
3: **procedure** AUTOMATON CONSTRUCTION
4:     Register the predicates appearing in the query: $a_1, \ldots, a_r$.
5:     Set the alphabet Σ of A equal to $\{a_1, \ldots, a_r\}$.
6:     Register the nodes of G that have an incoming or outgoing transition labeled
    with $a \in \Sigma$: $v_1, \ldots, v_n$.
7:     Set the states Q of A equal to $\{v_1, \ldots, v_n\}$.
8:     For each $a \in \Sigma$ construct the corresponding probability transition matrix M_a.
9:     Associate to each initial node v the appropriate unit vector $\mathbf{q}_{v_i}$.
10:     Associate to each final node v the appropriate unit vector $\mathbf{q}_{v_f}$.
11: **end procedure**
12: ◁ INPUT: A probabilistic automaton A and a threshold p ▷
13: ◁ OUTPUT: The answer to the query ▷
14: **procedure** COMPUTE QUERY ANSWER
15:     **for** each $a \in \Sigma$ **do**
16:         **while** $M_a^k \mathbf{q}_{v_i} > 0$ or $\mathbf{q}_{v_f}^T M_a^k > 0$ **do**
17:             Set $\mathbf{p}_{a,v_i}^k = M_a^k \mathbf{q}_{v_i}$.
18:             Set $(\mathbf{p}_{a,v_f}^k)^T = \mathbf{q}_{v_f}^T M_a^k$.
19:         **end while**
20:         Set $\mathbf{p}_{a,v_i} = \sum \mathbf{p}_{a,v_i}^k$.
21:         Set $(\mathbf{p}_{a,v_f})^T = \sum (\mathbf{p}_{a,v_f}^k)^T$.
22:     **end for**
23:     Compute the probability vector $\mathbf{p}$ corresponding to the query
24:     **for** each $v \in Q$ **do**
25:         **if** the corresponding entry in $\mathbf{p}$ is $\geq p$ **then** output v.
26:         **end if**
27:     **end for**
28: **end procedure**

Fig. 5. The general method for constructing probabilistic query automata.

5   Conclusion and Future Work

Semantic Web describes a notion of web content which is enriched with meaningful semantics. Linked Data and the RDF framework are the main mechanisms that are used for management of "smart" data that describes various entities and their relationships. On the other hand, SPARQL is the prevailing query language for information retrieval in semantic environments like Linked Data. SPARQL is currently capable of simple graph pattern queries, as well as queries for paths with transitive labels. Transitive predicates are met in multiple real-world domains, since transitivity is an inherent characteristic for the description of particular relations among objects.

Querying Linked Data can be achieved in various ways, such as link traversal based approaches, etc. Given the wealth and diversity of data currently available (and constantly increasing), the need for establishing appropriate methods and tools for querying is ever increasing. In many cases, the data described in such frameworks contains inherent notions of uncertainty. Biological and noisy data are important real-world examples. This parameter can be considered as the probability, or weight or degree of certainty.

While SPARQL does not currently support it, research efforts give us reason to expect that SPARQL will soon be extended with mechanisms that will allow the querying of probabilistic Linked Data. Towards this goal, we present the first to our knowledge approach that employs probabilistic automata for this particular task. This approach inherits all the advantages of simple computation models, like automata and can easily be adapted to various settings.

In conclusion, we have used probabilistic automata and we demonstrated that they can be associated with queries on RDF triples, in the presence of uncertainty denoted by a value attached to each triple. Future directions include expanding the model to handle potentially infinite uncertain data. In this case, the standard finite probabilistic automata may not be adequate and an infinite extension, like probabilistic ω-automata could be considered [3]. Such an approach would leverage the expressive power and the decidability results of the probabilistic version of ω-automata to enhance query response and process time.

References

1. SPARQL 1.1 Query Language. Technical report, W3C (2013), http://www.w3.org/TR/sparql11-query
2. Akbarinia, R., Valduriez, P., Verger, G.: Efficient evaluation of SUM queries over probabilistic data. IEEE Trans. Knowl. Data Eng. **25**(4), 764–775 (2013)
3. Baier, C., Grösser, M., Bertrand, N.: Probabilistic ω-automata. J. ACM **59**(1), 1–52 (2012)
4. Barceló, P., Libkin, L., Reutter, J.L.: Querying regular graph patterns. J. ACM (JACM) **61**(1), 8 (2014)
5. Dalvi, N., Suciu, D.: Efficient query evaluation on probabilistic databases. The VLDB J.- Int. J. Very Large Data Bases **16**(4), 523–544 (2007)
6. Fang, H., Zhang, X.: pSPARQL: a querying language for probabilistic RDF. In: Proceedings of ISWC Posters and Demos (2016)
7. Fernandez, M., Suciu, D.: Optimizing regular path expressions using graph schemas. In: Proceedings of the 14th International Conference on Data Engineering, pp. 14–23. IEEE (1998)
8. Giannakis, K., Andronikos, T.: Querying linked data and Büchi automata. In: 2014 9th International Workshop on Semantic and Social Media Adaptation and Personalization (SMAP), pp. 110–114. IEEE (2014)
9. Giannakis, K., Theocharopoulou, G., Papalitsas, C., Andronikos, T., Vlamos, P.: Associating ω-automata to path queries on Webs of Linked Data. Eng. Appl. Artif. Intell. **51**, 115–123 (2016)
10. Hartig, O.: An overview on execution strategies for Linked Data queries. Datenbank-Spektrum **13**(2), 89–99 (2013)
11. Hua, M., Pei, J.: Probabilistic path queries in road networks: traffic uncertainty aware path selection. In: Proceedings of the 13th International Conference on Extending Database Technology, pp. 347–358. ACM (2010)

12. Huang, H., Liu, C.: Query evaluation on probabilistic RDF databases. In: Vossen, G., Long, D.D.E., Yu, J.X. (eds.) WISE 2009. LNCS, vol. 5802, pp. 307–320. Springer, Heidelberg (2009). https://doi.org/10.1007/978-3-642-04409-0_32
13. Khan, A., Chen, L.: On uncertain graphs modeling and queries. Proc. VLDB Endowment 8(12), 2042–2043 (2015)
14. Krompaß, D., Nickel, M., Tresp, V.: Querying factorized probabilistic triple databases. In: Mika, P., et al. (eds.) ISWC 2014. LNCS, vol. 8797, pp. 114–129. Springer, Cham (2014). https://doi.org/10.1007/978-3-319-11915-1_8
15. Lian, X., Chen, L., Wang, G.: Quality-aware subgraph matching over inconsistent probabilistic graph databases. IEEE Trans. Knowl. Data Eng. 28(6), 1560–1574 (2016)
16. Marshall, M.S., Boyce, R., Deus, H.F., Zhao, J., Willighagen, E.L., Samwald, M., Pichler, E., Hajagos, J., Prud'hommeaux, E., Stephens, S.: Emerging practices for mapping and linking life sciences data using RDF-a case series. Web Semant. Sci. Serv. Agents World Wide Web 14, 2–13 (2012)
17. Paz, A.: Introduction to probabilistic automata. Academic Press Inc., Orlando (1971)
18. Rabin, M.O.: Probabilistic automata. Inf. Control 6(3), 230–245 (1963)
19. Reynolds, D.: Position paper: uncertainty reasoning for linked data. In: Workshop, vol. 14 (2014)
20. Schoenfisch, J.: Querying probabilistic ontologies with SPARQL. In: Proceedings GI-Edition, vol. 232, pp. 2245–2256 (2014)
21. Sistla, A.P., Hu, T., Chowdhry, V.: Similarity based retrieval from sequence databases using automata as queries. In: Proceedings of the Eleventh International Conference on Information and Knowledge Management, pp. 237–244. ACM (2002)
22. Theocharopoulou, G., Giannakis, K.: Web mining to create semantic content: a case study for the environment. In: Iliadis, L., Maglogiannis, I., Papadopoulos, H., Karatzas, K., Sioutas, S. (eds.) AIAI 2012. IFIP AICT, vol. 382, pp. 411–420. Springer, Heidelberg (2012). https://doi.org/10.1007/978-3-642-33412-2_42
23. Wang, X., Ling, J., Wang, J., Wang, K., Feng, Z.: Answering provenance-aware regular path queries on RDF graphs using an automata-based algorithm. In: Proceedings of the 23rd International Conference on World Wide Web, pp. 395–396. ACM (2014)
24. Zhang, X., Feng, Z., Wang, X., Rao, G., Wu, W.: Context-free path queries on RDF graphs. In: Groth, P., Simperl, E., Gray, A., Sabou, M., Krötzsch, M., Lecue, F., Flöck, F., Gil, Y. (eds.) ISWC 2016. LNCS, vol. 9981, pp. 632–648. Springer, Cham (2016). https://doi.org/10.1007/978-3-319-46523-4_38

Scaling and Cost Models in the Cloud

Improving Rule-Based Elasticity Control by Adapting the Sensitivity of the Auto-Scaling Decision Timeframe

Demetris Trihinas$^{(\boxtimes)}$ (iD), Zacharias Georgiou, George Pallis, and Marios D. Dikaiakos

Department of Computer Science, University of Cyprus, Nicosia, Cyprus
{trihinas,zgeorg03,gpallis,mdd}@cs.ucy.ac.cy

Abstract. Cloud computing offers the opportunity to improve efficiency with cloud providers offering consumers the ability to automatically scale their applications to meet exact demands. However, "auto-scaling" is usually provided to consumers in the form of metric threshold rules which are not capable of determining whether a scaling alert is issued due to an actual change in the demand of the application or due to short-lived bursts evident in monitoring data. The latter, can lead to unjustified scaling actions and thus, significant costs. In this paper, we introduce AdaFrame, a novel library which supports the decision-making of rule-based elasticity controllers to timely detect actual runtime changes in the monitorable load of cloud services. Results on real-life testbeds deployed on AWS, show that AdaFrame is able to correctly identify scaling actions and in contrast to the AWS auto-scaler, is able to lower detection delay by at least 63%.

Keywords: Cloud computing · Auto-scaling · Elasticity
Cloud monitoring

1 Introduction

Cloud computing is dominating the interests of multiple business domains revolutionizing the IT industry to the point where any person, with even basic technical skills, can access via the internet, vast and scalable computing resources by shifting IT spending to a pay-as-you-use model [1]. For small businesses and startups, this well-established argument is sound. Cloud computing eliminates capital expense of buying hardware and diminishes costs for configuring and running on-site computing infrastructures of any size [2]. Nevertheless, driving cloud adoption is *elasticity*, that is the ability of cloud services to acquire and release dedicated resources to meet current demand [3].

Albeit, while elasticity is one of cloud computing most-hyped features, the reality does not necessarily measure up to cloud providers' promises. For instance, application traffic from sudden user demand can explode rapidly, and

© Springer International Publishing AG, part of Springer Nature 2018
D. Alistarh et al. (Eds.): ALGOCLOUD 2017, LNCS 10739, pp. 123–137, 2018.
https://doi.org/10.1007/978-3-319-74875-7_8

the need for immediate scalability to address demands comes with many impediments. Cloud providers, such as AWS, offer "auto-scaling" by automatically provisioning VMs when certain user-defined metric thresholds are violated. Metric thresholds are usually reactive and rule-based in the IF-THEN-ACTION format (e.g., IF cpuUsage >75% THEN addVM) [4]. However, auto-scaling is challenging, especially when determining whether a scaling alert is issued due to actual change in the demand of an application, or due to sudden and short-lived (e.g., few seconds) spikes on highly sensitive monitoring data (e.g., cpu usage). The latter may resort to "ping-pong" effects where resources are provisioned and de-provisioned rapidly, but most importantly are billed although real demand does not exist [5]. Thus, *rapid scaling could, in fact, end up being detrimental resulting in unwanted charges.* On the other hand, delaying to determine an actual change in the application monitorable load by extending the ruling to include a time window that the scaling alert must persist (e.g., AWS default timeframe is 5 min), inhibits the possibility of a severe performance penalty affecting the overall application quality-of-service. In contrast to rule-based auto-scaling, a number of interesting and more advanced approaches have been proposed to offer better elasticity control based on machine learning and control theory [6–9]. However, cloud providers, for the time being, refrain from embracing such approaches as they suffer from practical limitations that derive from the complexity of the algorithmic process in a fully automated environment or the assumption that users have a priori knowledge of optimal parameter configuration.

In this paper, we introduce AdaFrame, a library for cloud provider rule-based and reactive auto-scalers to improve elasticity control, by supporting the decision-making process to timely detect actual runtime changes in the statistical properties of the monitorable load of cloud services. To achieve this, our library employs an online, low-cost and probabilistic algorithmic process based on runtime change detection which allows for elasticity controllers to reduce the possibility of falling victims to ping-pong effects without the need to resort to large decision timeframes which inquire significant performance penalties to cloud applications and their owners. In our evaluation with two real-world testbeds deployed on AWS, we show that AdaFrame is able to correctly notify of when scaling actions should be executed, and in comparison to AWS auto-scaling, AdaFrame is able to lower detection delay by at least 63%.

The rest of this paper is structured as follows: Sect. 2 elaborates the motivation. Section 3 introduces our library and the algorithmic process of the approach. Section 4 presents a comprehensive evaluation in real-life settings, while Sect. 5 presents the related work. Finally, Sect. 6 conclude this paper.

2   Motivation

Fostered by autonomic computing concepts, "auto-scaling" is now a fundamental process for market leading cloud service providers. This is commonly implemented as a decision-making problem, where resource allocation for an application consists of periodically monitoring the application load, the current allocated resources (e.g., number of VMs) and based on some scaling policy, decide to (de-)allocate resources in order to maintain the performance as close as possible to a target performance. Rule-based scaling policies are very popular among cloud providers and their consumers as the simplicity and intuitive nature of these policies make them very appealing.

These scaling policies are expressed with **IF <Expr> THEN <Action>** rules. In particular, consumers are usually not restricted to the number of rules they can define, while each rule is comprised of an expression (`<Expr>`) and scaling action (`<Action>`). The expression defines the target metric of interest and the relation which will trigger the scaling policy. Triggering the scaling policy will satisfy the desired action which is pre-selected by the consumer from a finite set of permitted scaling actions supported by the cloud provider. For example, let us consider a web service processing requests for local business outlets in the location defined as a parameter in the served request. In this scenario the load is compute-bound and we assume that two scaling policies are defined. The first policy, if triggered, will add a new virtual instance to the deployed cluster when the average cluster CPU utilization surpasses 80%, while the second policy, will remove a virtual instance if CPU utilization drops below 20%.

```
RULE$1 := IF AVG(cpuUsage(clusterID)) > 80% THEN addVM
RULE$2 := IF AVG(cpuUsage(clusterID)) < 20% THEN removeVM
```

While the simplicity is highly evident from the above example (although selecting appropriate thresholds is a profiling challenge of its own [6]) this approach ignores the volatility of monitoring data. To be precise, monitoring data can be bursty introducing sudden and short-lived spikes which may cause control oscillations. A control oscillation, often dubbed as *a "ping-pong" effect, is defined as the phenomenon where an unexpected and short-lived burst in the monitoring data (even a single datapoint) triggers a scaling action which will be subsequently*

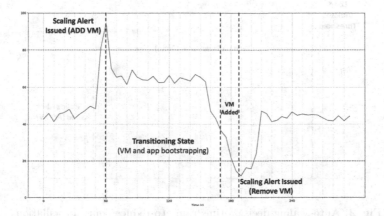

Fig. 1. Ping-Pong effect on monitored CPU utilization of a web service

annulled when the system stabilizes [10]. As an illustrating example, let us consider the aforementioned web service with its CPU usage depicted in Fig. 1. From this, we observe that a sudden burst from a background cleanup process immediately triggers RULE\$1, thus a VM is added, and after the VM is provisioned and fully integrated to the deployment, the CPU utilization, quickly, drops to the point where RULE\$2 is triggered, returning the application deployment to the previous state. Hence, a single spike in the monitoring data causes a series of elastic control actions, accounting for direct and indirect costs, as users are charged for these unjustified actions and the provisioned resources (e.g., for AWS a VM booted even for 1 s is charged for whole hour), and the application may suffer performance-wise if data movement and coordination is required while in a transitioning state.

To compensate with control oscillations, cloud providers (e.g., AWS), extend the rule-based decision model to include a time window, denoted as a decision timeframe, where the scaling policy expression (<Expr>) must evaluate to true and persist for the length of the timeframe. In particular, AWS, provides its consumers with the option to set a decision timeframe, with the default option being 5 min, while other options are also available [11]. The assumption here is that if a threshold violation persists in time, as depicted in Fig. 2, then a scaling action is justified and the larger the decision timeframe, the smaller the possibility of introducing a ping-pong effect. Obviously, absolute guarantees can never be given unless auto-scaling is disabled. However, the downside with introducing a large decision timeframe, even the default AWS option of 5 min, is that a significant performance penalty may occur while waiting for a scaling action to be triggered. Therefore, a new challenge rises: *How can a rule-based elasticity controller scale an application deployment without resorting to large decision timeframes in order to avoid ping-pong effects due to sudden bursts in monitoring data?*

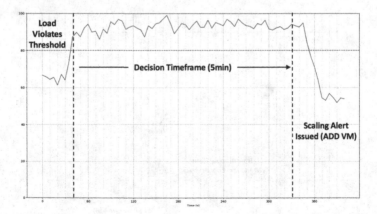

Fig. 2. Auto-scaling decision timeframe to reduce control oscillations

3   The AdaFrame Library

To address the aforementioned challenge, we have designed the AdaFrame library. AdaFrame supports the decision-making process of rule-based elasticity controllers so as to timely detect, and notify the elasticity controller, of actual runtime changes in the statistical properties of monitoring data originating from elastic cloud services. To achieve this, the AdaFrame library employs an online, low-cost and probabilistic algorithmic process based on change detection which allows for elasticity controllers to reduce the possibility of falling victims to ping-pong effects without the need to resort to large decision timeframes which inquire significant performance penalties to cloud applications and their owners. Figure 3 depicts the AdaFrame library incorporated in an auto-scaling control loop, resembling AWS, where one can observe that it does not alter the decision-making process, or the control loop in general, as AdaFrame simply acts as a support proxy notifying the Scaling Policy Evaluation of the elasticity controller when a scaling action should be triggered (workload behavior changes) and when not (workload spikes). This completely removes the need of a fixed decision timeframe. In turn, offline profiling to detect a (near-) optimal decision timeframe is not needed, as fixed "optimal" values are only relevant if the properties of the metric stream hold for the entire lifespan of the application which is an assumption far from reality for today's complex cloud applications. In the following, we provide a detail description of the two basic components comprising AdaFrame: the Adaptive Monitoring Estimation Model and Runtime Change Detection.

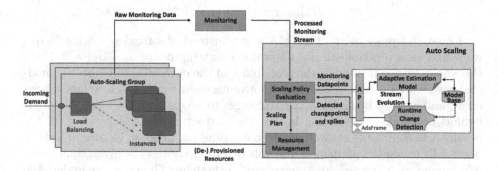

Fig. 3. AdaFrame incorporated in auto-scaling process

3.1   Adaptive Monitoring Estimation Model

At first, let us define a monitoring stream $M = \{d_i\}_{i=0}^{n}$ published by a monitoring source on a cloud application to an auto-scaling entity, as a large sequence of datapoints d_i, where $i = 0, 1, ..., n$ and $n \to \infty$. Each datapoint d_i is a tuple (s_{id}, t_i, v_i) described, at the minimum, by a source identifier s_{id}, a timestamp t_i and a value v_i. We base our approach such that the estimation model is maintained in constant time and space O(1), a requirement for rule-based elasticity control. While AdaFrame supports model parameterization, as input it only requires from the user to provide his/her confidence guarantees $\delta \in [0, 1]$, denoting the probability with which estimated datapoints are approximated from sensed datapoints. Now, when a datapoint is made available to the auto-scaling by the monitoring tool, it is passed through `AdaFrame API` to the `Adaptive Monitoring Estimation Model` so as to update the current monitoring stream evolution by using a moving average, denoted as μ_i. This will give an initial estimation for the next datapoint value, denoted as $\hat{v}_{i+1}$. Moving averages provide smoothing and one-step ahead estimations for single dimensional timeseries such as the target metric referenced in the `<Expr>` of a rule-based scaling policy. They are easy to compute, though many types exist, and can be calculated on the fly with only previous value knowledge. A cumulative moving average for streaming data is the Exponential Weighted Moving Average (EWMA), $\mu_i = \alpha\mu_{i-1} + (1 - \alpha)v_i$, where a weighting parameter α, is introduced to decrease exponentially the effect of older values. However, the EWMA features a significant drawback; it is volatile to abrupt transient changes [12]. Thus, we propose adopting a Probabilistic EWMA (PEWMA), which dynamically adjusts the weighting based on the probability density of the given observation. The PEWMA acknowledges sufficiently abrupt transient changes (burstiness), adjusting quickly to long-term shifts in the monitoring stream evolution and when incorporated in our algorithmic estimation process, it requires no parameterization, scaling to numerous datapoints.

$$\mu_i = \begin{cases} v_i, & i = 1 \\ \alpha(1 - \beta P_i)\mu_{i-1} + (1 - \alpha(1 - \beta P_i))v_i, & i > 1 \end{cases} \tag{1}$$

Equation 1 presents the PEWMA where instead of a fixed weighting factor, we introduce a probabilistically adaptable weighting factor $\tilde{a}_i = \alpha(1 - \beta P_i)$. In this equation, the p-value, is the probability of the current v_i to follow the modeled distribution of the metric stream evolution. In turn, β is a weight placed on P_i and as $\beta \to 0$ the PEWMA converges to a common EWMA[1]. The logic behind probabilistic reasoning is that the current value v_i depending on it's p-value will contribute respectively to the estimation process. In turn, if a datapoint falls inside the prediction intervals determined from the given confidence, it is labeled as "expected" or "unexpected" otherwise. Therefore, we update the weighting by $1 - \beta P_i$ so that sudden "unexpected" spikes are accounted for in the estimation process, however, offer little influence to subsequent estimations, thus restraining the model from overestimating subsequent v_i's. In turn, if an "unexpected" value turns out to be a shift in the monitoring stream evolution, as the probability kernel shifts, subsequent "unexpected" values are awarded with greater p-values, allowing them to contribute more to the estimation process. Assuming, a stochastic and i.i.d distribution as the bare minimum for a monitoring stream, we can adopt a Gaussian kernel $N(\mu, \sigma^2)$, which satisfies the

---

[1] For simplicity in our model we will consider $\beta = 1$.

aforementioned requirements. Thus, P_i is the probability of v_i evaluated under a Gaussian distribution, which is computed by Eq. 2. Nonetheless, we note that while a Gaussian distribution is assumed, if prior knowledge of the distribution is available and given by the user then only the computation of P_i must change in the estimation process.

$$P_i = \frac{1}{\sqrt{2\pi}} \exp(-\frac{Z_i^2}{2})$$
$$Z_i = \frac{v_i - \hat{v}_i}{\sigma_i}$$

(2)

Moreover, in [12] we show how to compute the running variance for the PEWMA, and that the α parameter can take a wide range of values if a small imprecision can be tolerated as most of the error is absorbed by the probabilistic weighting. Thus, with the proposed model we can both estimate the monitoring stream evolution and detect and label bursts in the monitoring stream. However, as mentioned, much of this burstiness is irrelevant for diagnosis of elastic scaling. Nonetheless, significant bursts and long-term trends are useful features for cloud providers and consumers, especially for capacity planning, anomaly detection and quality control.

3.2   Runtime Change Detection

The most prominent functionality of AdaFrame is to detect changes, at runtime, in the statistical properties of the evolution of a monitoring stream, in order to reduce the need of a fixed decision timeframe to avoid ping-pong effects. To achieve this, AdaFrame incorporates change detection based on a variation of the lightweight Cumulative Sum test (CUSUM). The CUSUM, denoted as C_i, is a hypothesis test for detecting shifts in i.i.d timeseries [13]. In particular, there are two hypothesis θ' and θ'' with probabilities $P(M, \theta')$ and $P(M, \theta'')$, where the first corresponds to the statistical distribution of the monitoring stream prior to a shift ($i < t_s$) and the second to the distribution after a shift ($i > t_s$) with t_s denoting the time interval the shift/change occurs. The CUSUM is computed online via sequential probability testing on the instantaneous log-likelihood ratio given for a monitoring stream at the i-th time interval, as follows:

$$c_i = \ln \frac{P(M_i, \theta'')}{P(M_i, \theta')}$$
$$C_{i,\{low,\ high\}} = C_{i-1,\{low,\ high\}} + c_i$$

(3)

where low and $high$ denote the separation of the CUSUM to identify both positive and negative shifts respectively. The typical behavior of the log-likelihood ratio includes a negative drift before a shift and a positive drift after the shift. Thus, the relevant information for detecting a shift in the evolution of a monitoring stream lays in the difference between the value of the log-likelihood ratio and the current minimum value. A decision function G_i, is used to determine a shift in the monitoring stream when its outcome surpasses a threshold h, measured in standard deviation units. The time interval at which a shift actually occurs, is computed from the CUSUM as follows:

$$G_{i,\{low,\ high\}} = \{G_{i-1,\{low,\ high\}} + c_i\}^+$$
$$t_s = \arg\min_{j \le s \le i} (C_{s-1}) \tag{4}$$

In the above, $G^+ = sup(G, 0)$ and t_i is the time AdaFrame detects the shift. Now, let us consider the particular case of a monitoring stream representing the target metric of a rule-based scaling policy with the monitoring stream supposed to undergo possible shifts in its evolution. Hence, in our case, t_j is considered the time the monitoring stream current value base violates the scaling policy. In turn, we consider the evolution of the monitoring stream in its mean, modelled as the PEWMA moving average previously introduced. Thus, θ' and θ'' can be rewritten as μ' and μ'' respectively, with μ' representing the current evolution, while μ'' the output of the estimation model with $\mu'' = \mu' + \epsilon$, and ϵ denoting the estimated magnitude of change of the monitoring stream evolution. As the monitoring stream evolution is used to provide an estimation for $\hat{v}_i$, the magnitude of change is actually equal to $\epsilon = \hat{v}_i - v_i$. In turn, let $P(M, \mu')$ and $P(M, \mu'')$ be

---

Algorithm 1. AdaFrame Algorithm

Input: User-provided confidence δ_i at initialization. For every update, datapoint $d(t_i, v_i)$
Output: Label datapoint d_i as "expected", "unexpected" or "changepoint"
Ensure: Monitoring stream M is attached and moving average μ is initialized

    compute p– and z–value and then update estimation model
1: $P_i, Z_i \leftarrow$ probDistro$(v_i, \hat{v}_i, \sigma_i)$ (Eq. 2)
2: $\mu_i, \sigma_i \leftarrow$ updPEWMA(P_i, v_i) (Eq. 1)
    label datapoint as "expected" or "unexpected" based on prediction intervals
3: **if** isDatapointExpected(δ, P_i, Z_i) **then**
4:     $label \leftarrow$ 'expected'
5: **else**
6:     $label \leftarrow$ 'unexpected'
7: **end if**
8: **if** *scaling alert triggered at* t_{i-1} **then**
9:     $h_i \leftarrow$ updShiftThres(δ, σ_i) (Eq. 6)
10: **end if**
11: $c_i \leftarrow$ updLikelihood$(v_i, \hat{v}_i, \mu_i, \sigma_i)$ (Eq. 5)
12: $C_{i,low}, C_{i,high} \leftarrow$ updCusum(c_i) (Eq. 3)
13: $G_{i,low}, G_{i,high} \leftarrow$ updDecision(c_i) (Eq. 4)
14: **if** $G_{i,\{low,\ high\}} > h_i$ **then**
15:     $label \leftarrow$ 'changepoint'
16: **end if**
17: **return** *label*

---

computed from Eq. 2. With some calculations, c_i (Eq. 4) is rewritten, as follows, to perform the decision-making with only previous value knowledge:

$$c_{i,\{low,\ high\}} = \pm \frac{|\epsilon|}{\sigma_i^2} \left(v_i - \mu' \mp \frac{|\epsilon|}{2}\right) \tag{5}$$

Nonetheless, the CUSUM test features two drawbacks. First, determining the actual t_s requires linear time. However, exact knowledge of t_s is not required for signalling a scaling action, as t_s is only computed after the shift is detected, with AdaFrame providing an approximate answer (t_i) which is the time it detects the change in the monitoring stream. Second, when the monitoring stream is relatively stable, and thus the stream variance is low $(\sigma_i \to 0)$, the CUSUM is prone to falsely signalling changes [14]. Hence, we follow an adaptive approach where h is updated after a scaling action, based on the number of standard deviations respecting the given user-defined confidence (δ) and an optional positive value (h_{min}) is used to restrict the sensitivity of the CUSUM so as to not oscillate between low values when the monitoring stream is relatively stable.

$$h_i = \max\{h_{min},\ h(\delta)\} \tag{6}$$

4 Evaluation

In this section, we evaluate the accuracy our approach via two real-world testbeds deployed on AWS, the most popular cloud provider. We compare AdaFrame ability to detect changes in the workload and signal the auto-scaling service that a scaling action should be triggered, to (i) using AWS auto-scaling without a decision timeframe, and, thus, any scaling policy violation will trigger a scaling action; and (ii) the AWS auto-scaling service with the default decision timeframe of 5 min.

To integrate AdaFrame to the AWS control loop, we enabled manual auto-scaling through the AWS API. This provides us with API access to the auto-scaling service so as to immediately trigger the pre-selected scaling action once a changepoint is detected after the initial scaling policy violation occurs. We note that when running AdaFrame with "manual" scaling, auto-scaling, with 5 min decision timeframe, is also enabled to see if any scaling action would be detected earlier by AWS. Also, both experiments are conducted with a tight confidence parameter of $\delta = 0.95$ and $h_{min} = 1$. In turn, as AWS monitoring metric collection limits the minimum periodicity to 60 s, for the sake of a thorough evaluation we opted to integrate with AWS, JCatascopia monitoring probes [10] in order to collect data every 5 s.

4.1 Testbed 1: Scaling a NoSQL Document Store

The first testbed of our evaluation is a NoSQL document store implemented by a Couchbase DB cluster. In particular, couchbase is used for the database back-end of the web service described in Sect. 2, and thus, processes map/reduce-like data requests for local business outlets in the location defined as a parameter in the served request. Initially, we manually provision the cluster to host three database instances which is considered, for Couchbase, as the minimum number of instances for smooth operation. Each instance, and future provisioned instances, are Amazon ubuntu 16.04 LTS medium flavored AMIs (2 VCPU, 4 GB Memory, 120 GB disk). For this testbed, and with Couchbase cpu-bound, we select the average cpu usage as the target metric. We stress the testbed by generating a stable load of 80 req/s and increase the request rate by 30 req/s every 10 min in order for the testbed to scale but not be overwhelmed while the size of the testbed is small. To cope with the workload, the AWS auto-scaling service must provision a new VM, in the first case, if the CPU usage is over 75% and, in the second case, if the same scaling policy is violated but for a timespan of 5 min. In the case of embracing AdaFrame, a new instance is only provisioned when a changepoint is detected after the scaling policy is violated.

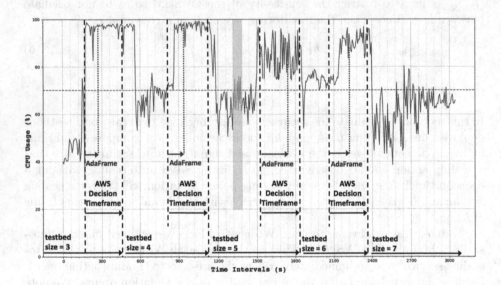

Fig. 4. Couchbase CPU usage and scaling action detection delay

Figure 4 depicts the testbed CPU usage, the testbed size, the time intervals at which the scaling policy is violated for the first time and the time intervals AWS (with decision timeframe) and AdaFrame trigger each scaling action. From this, we immediately observe that *AdaFrame features the ability to correctly identifying all scaling actions, even for a monitoring stream featuring significant burstiness over the threshold, and does not trigger any false scaling action which could lead to a ping-pong effect*. Also, scaling action detection is performed by AdaFrame in a timely manner (Fig. 8). Specifically, in three out of four of the scaling actions, the detection time is significantly less that half (63% less) of the AWS decision timeframe (112 s $\pm$ 16 s) as AdaFrame quickly identifies a change in the statistical properties of the monitoring stream. Nonetheless, for the third action, due to the high volatility of the monitoring stream, AdaFrame requires more than half of the decision timeframe (196 s), but still significantly outperforms AWS (36% less). In turn, the highlighted period of time in Fig. 4 is an example where even if a smaller, but still fixed, decision timeframe (e.g., 2 min) is opted for AWS to compete with AdaFrame, then a ping-pong effect will be triggered as oscillations near the threshold are evident, which justifies why ignoring the statistical properties of the monitoring stream to detect actual changes will lead to unexpected and unwanted effects.

Next, to illustrate the importance of timely detecting when to trigger a scaling action, we depict in Fig. 5, the performance of the testbed in terms of throughput when a decision timeframe is used. From this, we observe that once a scaling policy violation is detected and for the span of the decision timeframe, throughput suffers and is not able to follow the workload increment. Only after a VM is added and aftergoing a (short) rebalancing phase is the testbed able to surpass the initial saturation. In contrast, *AdaFrame is able to correctly and timely detect when a scaling action should be triggered (Fig. 5), and we observe that throughput features a significantly larger slop and higher values are achieved with the gains increasing as the workload increases (Fig. 6).* Finally, we note that for visualization clarity, we omitted depicting cpu usage, throughput and time intervals at which AWS without a decision timeframe triggers a scaling action, and state that in this case correctness suffers as 7 scaling actions were triggered instead of 4 due to the high volatility of the monitoring stream near the threshold.

4.2   Testbed 2: Scaling the Business Logic of a Web Service

The second testbed of our evaluation is an Apache Tomcat cluster implementing the business logic of the aforementioned web service. This cluster was initially

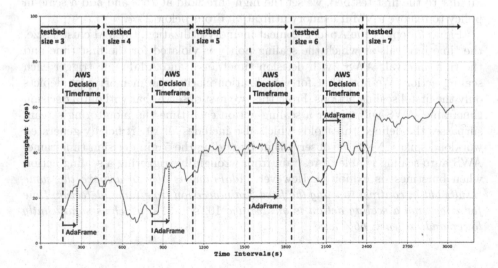

Fig. 5. Couchbase cluster throughput and scaling action detection delay

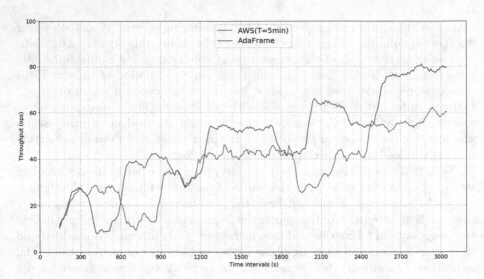

Fig. 6. Couchbase cluster throughput with AdaFrame vs 5 min decision timeframe

provisioned to host a single instance and each provisioned instance, is an Amazon ubuntu 16.04 LTS small flavored AMI (1 VCPU, 2 GB Memory). In this set of experiments, we configure our workload generator to adapt the load step-size randomly in order to cause both scale in and out actions. The number of scaling actions to perform was set to 10. Also, we configure the workload generator, to exhibit at random, 10 periods of time with bursty behavior over the defined thresholds by mixing with the workload gaussian noise to emulate the spiky behavior of a (Tomcat) cleanup background process. For this testbed, and with Apache Tomcat memory-bound, we select the average memory utilization as the targeted metric. Similar to the first testbed, we set the high threshold at 75% and add a scale-in policy to remove a VM if memory utilization drops below 25%.

Figure 7 depicts the Apache Tomcat memory utilization, the VM cluster size, the time intervals at which the scaling policy is violated for the first time and the time intervals AWS (with decision timeframe) and AdaFrame trigger each scaling action. We note that for visualization clarity, the memory plot depicts only the first 4 scaling actions. From this, we first observe that without a decision timeframe, AWS will trigger a scaling action each time the monitoring stream surpasses the defined thresholds which also includes all 10 artificially generated workload spikes. Next, we observe that by adding the 5 min decision timeframe, AWS auto-scaling is able to restrain from wrongfully triggering a scaling action when burstiness is exhibited. However, *AdaFrame is able to achieve the same results but in contrast to using the fixed 5 min decision timeframe, detection time for triggering a scaling action is on average 102 s ± 21 s, which is significantly lower and, at least, 66% less.*

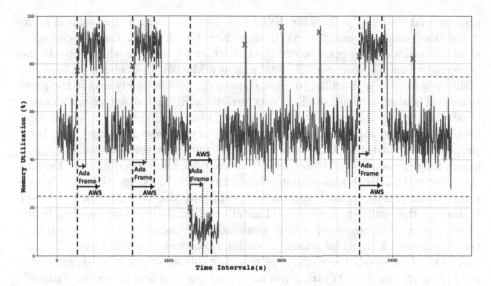

Fig. 7. Apache Tomcat cluster memory utilization and scaling action detection delay

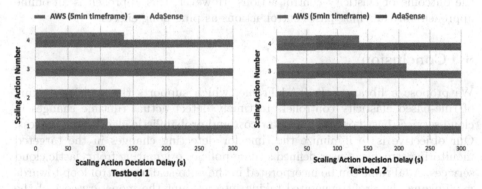

Fig. 8. Scaling action decision delay per testbed

5   Related Work

A number of sophisticated techniques have been proposed for elastic scaling. Almeida et al. [8] propose a *branch and bound* approach for optimally allocating resources to multi-layer cloud applications during runtime, while Tolosana-Calasanz et al. [9] propose controlling reserved resources for data processing engines by following a shared token bucket approach. A more intuitive approach is proposed by Dustdar et al. [3], defining elasticity as a complex property, having as major dimensions resource, cost and quality elasticity. These dimensions reflect not only computing related aspects of application operation, but also business aspects. In turn, Copil et al. [15] introduce an elasticity specification language, denoted as SYBL, which allows the definition of complex and multidimensional elasticity policies for rSYBL, an elasticity controller capable of managing cloud elasticity based on SYBL directives. On the other hand, Tsoumakos et al. [6] introduce an open-source elasticity control service, which models the problem of elastic scaling NoSQL databases as a Markovian Decision Process and utilize reinforcement learning to allow the system to adaptively decide the most beneficial scaling action based on user policies and past observations. Naskos et al. [16] extend this model to resizing clusters of a single generic application hosted on virtual machines. Many queuing theory based approaches have been proposed. For instance, Urgaonkar et al. [17] models servers at each tier as a queue for representing arbitrary arrival and service time distributions. Despite the novelty in all the above approaches, cloud providers, for the time being, refrain from embracing such approaches, and prefer the simplicity of rule-based scaling, as they suffer from practical limitations that derive from the complexity of the algorithmic process in a fully automated environment or the assumption that users have a priori knowledge of optimal parameter configuration. Finally, and to the best of our knowledge, the most notable approach towards attacking ping-pong effects, is ADVISE, a framework supporting and providing "advise" to elasticity controllers to improve the decision-making process by evaluating the outcome of elasticity control actions. However, this approach is an offline approach only labelling past control actions as ping-pong effects.

6   Conclusion

We propose a library, called AdaFrame, which supports the decision-making of rule-based elasticity controllers to timely detect actual runtime changes in cloud services based on an online, low-cost and probabilistic algorithmic process. Our objective is to minimize the time for detecting changes in the targeted monitoring streams of user-defined scaling policies originating from elastic cloud services. AdaFrame can be incorporated in the auto-scaling control loop towards maximizing the profit generated taking into account the monetary cost of the resources as well as the revenue generated by the workload. Results on two real-life testbeds deployed on AWS show that AdaFrame outperforms the AWS auto-scaler and adapts quickly to workload changes.

Acknowledgements. This work is partially supported by the European Commission in terms of Unicorn 731846 H2020 project (H2020-ICT-2016-1).

References

1. Loulloudes, N., Sofokleous, C., Trihinas, D., Dikaiakos, M.D., Pallis, G.: Enabling interoperable cloud application management through an open source ecosystem. IEEE Internet Comput. **19**(3), 54–59 (2015)
2. Willcocks, L., Venters, W., Whitley, E.A.: Cloud in context: managing new waves of power. In: Moving to the Cloud Corporation, pp. 1–19. Palgrave Macmillan, London (2014). https://doi.org/10.1057/9781137347473_1

3. Dustdar, S., Guo, Y., Satzger, B., Truong, H.L.: Principles of elastic processes. IEEE Internet Comput. **15**(5), 66–71 (2011)
4. Trihinas, D., Sofokleous, C., Loulloudes, N., Foudoulis, A., Pallis, G., Dikaiakos, M.D.: Managing and monitoring elastic cloud applications. In: Casteleyn, S., Rossi, G., Winckler, M. (eds.) ICWE 2014. LNCS, vol. 8541, pp. 523–527. Springer, Cham (2014). https://doi.org/10.1007/978-3-319-08245-5_42
5. Copil, G., Trihinas, D., Truong, H., Moldovan, D., Pallis, G., Dustdar, S., Dikaiakos, M.D.: Evaluating cloud service elasticity behavior. Int. J. Coop. Inf. Syst. (2015)
6. Tsoumakos, D., Konstantinou, I., Boumpouka, C., Sioutas, S., Koziris, N.: Automated, elastic resource provisioning for NoSQL clusters using TIRAMOLA. In: IEEE International Symposium on Cluster Computing and the Grid, pp. 34–41 (2013)
7. Lolos, K., Konstantinou, I., Kantere, V., Koziris, N.: Elastic resource management with adaptive state space partitioning of Markov Decision Processes. CoRR abs/1702.02978 (2017)
8. Almeida, A., Dantas, F., Cavalcante, E., Batista, T.: A branch-and-bound algorithm for autonomic adaptation of multi-cloud applications. In: 2014 14th IEEE/ACM International Symposium on Cluster, Cloud and Grid Computing, pp. 315–323, May 2014
9. Tolosana-Calasanz, R., Ángel Bañares, J., Pham, C., Rana, O.F.: Resource management for bursty streams on multi-tenancy cloud environments. Future Gener. Comput. Syst. **55**, 444–459 (2016)
10. Trihinas, D., Pallis, G., Dikaiakos, M.D.: Monitoring elastically adaptive multi-cloud services. IEEE Trans. Cloud Comput. **4** (2016)
11. Amazon Auto-Scaling Policies. http://aws.amazon.com/ec2/
12. Trihinas, D., Pallis, G., Dikaiakos, M.D.: AdaM: an adaptive monitoring framework for sampling and filtering on IoT devices. In: IEEE International Conference on Big Data, pp. 717–726 (2015)
13. Luo, Y., Li, Z., Wang, Z.: Adaptive cusum control chart with variable sampling intervals. Comput. Stat. Data Anal. **53**(7), 2693–2701 (2009)
14. Trihinas, D., Pallis, G., Dikaiakos, M.: ADMin: adaptive monitoring dissemination for the internet of things. In: IEEE INFOCOM 2017 - IEEE Conference on Computer Communications (INFOCOM 2017), Atlanta, USA, May 2017
15. Copil, G., Moldovan, D., Truong, H.L., Dustdar, S.: SYBL: an extensible language for controlling elasticity in cloud applications. In: 13th IEEE/ACM International Symposium on Cluster, Cloud and Grid Computing, pp. 112–119 (2013)
16. Naskos, A., Stachtiari, E., Gounaris, A., Katsaros, P., Tsoumakos, D., Konstantinou, I., Sioutas, S.: Dependable horizontal scaling based on probabilistic model checking. In: 2015 15th IEEE/ACM International Symposium on Cluster, Cloud and Grid Computing, pp. 31–40, May 2015
17. Urgaonkar, B., Shenoy, P., Chandra, A., Goyal, P., Wood, T.: Agile dynamic provisioning of multi-tier internet applications. ACM Trans. Auton. Adapt. Syst. **3**(1), 1:1–1:39 (2008)

Risk Aware Stochastic Placement of Cloud Services: The Multiple Data Center Case

Galia Shabtai[1]([✉]), Danny Raz[2], and Yuval Shavitt[1]

[1] School of Electrical Engineering, Tel Aviv University, Ramat Aviv,
69978 Tel Aviv, Israel
galiashabtai@gmail.com, shavitt@eng.tau.ac.il
[2] Faculty of Computer Science, The Technion, 32000 Haifa, Israel
danny@cs.technion.ac.il

Abstract. Allocating the right amount of resources to each service in any of the data centers in a cloud environment is a very difficult task. In a previous work we considered the case where only two data centers are available and proposed a stochastic based placement algorithm to find a solution that minimizes the expected total cost of ownership. This approximation algorithm seems to work well for a very large family of overflow cost functions, which contains three functions that describe the most common practical situations. In this paper we generalize this work for arbitrary number of data centers and develop a generalized mechanism to assign services to data centers based on the available resources in each data center and the distribution of the demand for each service. We further show, using simulations based on synthetic data that the scheme performs very well on different service workloads.

1 Introduction

The recent rapid development of cloud technology gives rise to "many-and-diverse" services being deployed in datacenters across the world. The placement of services to the available datacenters in the cloud has a critical impact on the ability to provide a ubiquitous cost-effective high quality service. There are many challenges associated with optimal service placement due to the large scale of the problem, the need to obtain state information, and the geographical spreading of the datacenters and users.

One intriguing problem is the fact that resource requirement of services changes over time and is not fully known at the time of placement. Moreover, while the average demand may follow a clear daily pattern, the actual demand of a service at a specific time may vary considerably according to the stochastic nature of the demand. One way of addressing this important problem is over-provisioning, that is, allocating resources according to the peak demand. Clearly, this is not a cost effective approach and much of the resources are unused most

This paper was supported in part by the Neptune Consortium, Israel and by the Israeli Ministry of Science, Technology and Space.

© Springer International Publishing AG, part of Springer Nature 2018
D. Alistarh et al. (Eds.): ALGOCLOUD 2017, LNCS 10739, pp. 138–156, 2018.
https://doi.org/10.1007/978-3-319-74875-7_9

of the time. A more economical approach, relying on the stochastic nature of the demand, is to allocate just the right amount of resources and potentially use additional mechanisms (such as diverting the service request to a remote location or dynamically buying additional resources) in case of overflow situations where demand exceeds the capacity. Clearly, the cost during such (overflow) events is higher than the normal cost. Moreover, in many cases it is proportional to the amount of unavailable resources. Obviously, the quantitative way of modeling the cost of an overflow situation, considerably depends on the actions taken (or not taken) in such cases. For example one may want to minimize the probability of an overflow event, or to minimize the expected overflow of the demand. The challenge is therefore to find an optimal placement for a given cost function.

The simplest, non-trivial variant is when there are two data centers in the cloud. This case was studied by us [1] where we combined techniques from two different lines of work, stochastic resource allocation [2–5] and convex optimization [6,7], to give an efficient robust algorithm for approximating the optimal solution. In this paper we extend this work to the general case of arbitrary number of data centers.

Our algorithm for the general case with multiple data centers follows the intuition explained in [1] for two data centers: the algorithm is trying to allocate more spare capacity to the high risk services. In our case this intuition translates to "The double-sorting algorithm" which sorts the data centers by their capacity, the services by their risk (defined as the amount of variance per unit of expectation) and then allocates consecutive segments of the sorted list of services to the data centers, with lower risk services allocated to smaller capacity data centers.

A challenge is finding the right partition of the sorted list to k consecutive segments, where k is the number of data centers (bins). In the two data centers case (where $k = 2$) this is done by a brute force search over the n possible partition points, and gives an almost linear time algorithm. When there are more than two data centers ($k > 2$) the brute force algorithm has complexity of about n^{k-1} and is impractical. The main bulk of the paper is devoted to describing an efficient algorithm for this problem and proving its correctness and complexity.

As explained before we are interested in giving a *generic* solution to the problem that is applicable to a wide class of cost functions. In the two-bin case [1] we showed how to solve the approximation problem for any cost function that respects three natural properties (see Sect. 4). In Sect. 6 we develop a dynamic programming algorithm that solves the k-bin case in time of about $O(n^3)$ for all cost functions in the above class.

Admittedly, while $O(n^3)$ gives a polynomial time algorithm independent of k, it may still be too expensive to run in practice, and therefore we next concentrate on presenting an efficient algorithms for specific, natural cost functions. We focus on three natural cost functions that are used in practice, which are SP-MED (stochastic placement with minimum expected deviation), SP-MWOP (stochastic placement with minimum worst overflow probability) and SP-MOP (stochastic placement with minimum overflow probability).

In Sect. 7 we devise a novel algorithm, the *moving sticks algorithm*, that uses a completely different approach for the problem. We formally analyze the algorithm for the SP-MWOP cost function, and show that it approximates the optimal solution in almost linear time. We also devise a similar algorithm for the other two cost functions, but we do not give a formal proof of correctness.

We implemented our algorithms and evaluated their performance over synthetic data. Our results indicate that the new algorithms achieve a considerable gain when compared with commonly used naive solutions.

2   Problem Formulation

The input to the problem consists of k and n, specifying the number of bins and services, and integers $\{c_j\}_{j=1}^{k}$, specifying the bin capacities. We are also given a partial description of n *independent* random variables $X = (X^{(1)}, \ldots, X^{(n)})$. This partial description includes the mean $\mu^{(i)}$, variance $V^{(i)} = \mathbb{E}(|X^{(i)} - \mu^{(i)}|^2)$ and $\rho^{(i)} = \mathbb{E}(|X^{(i)} - \mu^{(i)}|^3)$ of each variable $X^{(i)}$ ($\rho^{(i)}$ are needed only for the error estimation of the reduction to the normal case). The output is a partition of $[n]$ to k disjoint sets $S = S_1, \ldots, S_k \subseteq [n]$, where S_j includes indices of services that needs to be allocated to bin j. Our goal is to find a partition minimizing a given cost function $D(S)$.

In [1] we used the Berry-Esseen theorem (which is a quantitative version of the central limit theorem) to bound the gap between the optimal solution under independent *arbitrary* input demand distributions and the optimal solution under independent *normal* input demand distributions. We showed that when the number of services is large and the first three moments of the services satisfy a mild condition, there is a reduction from the general case, where the independent $X^{(i)}$ are almost arbitrary, to the normal case. The same reduction also applies in our case (when $k > 2$). Therefore, in this work we assume w.l.o.g. that the independent input services are normally distributed and in this case we do not need $\rho^{(i)}$.

3   Summary of Our Results for Two Data Centers

In this section we present the main results of the normal two bin case, studied in [1], which are needed for the understanding of the $k > 2$ case analysis presented in the sequel.

The input to the problem is $c_1, c_2, \{\mu^{(i)}, V^{(i)}\}_{i=1}^{n}$ as before. Define $a^{(i)} = \frac{\mu^{(i)}}{\mu}$, $b^{(i)} = \frac{V^{(i)}}{V}$ and $P^{(i)} = (a^{(i)}, b^{(i)})$. If we split the services according to the partition $S = (S_1 = I, S_2 = [n] \setminus I)$, then the first bin is normally distributed with mean $\mu_1 = \mu \sum_{i \in I} a^{(i)}$ and variance $V_1 = V \sum_{i \in I} b^{(i)}$. Therefore, $(\frac{\mu_1}{\mu}, \frac{V_1}{V}) = P_I = \sum_{i \in I} P^{(i)}$. The point P_I also determines the mean and variance of the second bin: $(\frac{\mu_2}{\mu}, \frac{V_2}{V}) = (1, 1) - P_I$. Hence, we can associate the partition S with the point P_I.

In [1] we defined a function $D : [0,1]^2 \to \mathbf{R}$ where $D(a,b)$ is the cost function D under a partition where the demand to the first bin is normally distributed with mean $a\mu$ and variance bV. We also decoupled the optimization problem into two separate and almost orthogonal problems: the first is understanding the feasible set of solutions, and the second is the behavior of the objective cost function as a *continuous* function over the two-dimensional domain.

A point P_I that is associated with a partition $S = (I, [n] \setminus I)$ is called *an integral point*. The convex hull of all the integral points is the set of all fractional points. The first task is to find a convenient description of this convex set.

Definition 1 *(The sorted paths). Sort the services by their variance to mean ratio (VMR) in increasing order (i.e., $\frac{V^{(1)}}{\mu^{(1)}} \le \frac{V^{(2)}}{\mu^{(2)}} \le \cdots \le \frac{V^{(n)}}{\mu^{(n)}}$) and calculate the $P^{(1)}, P^{(2)}, \ldots, P^{(n)}$ vectors. For $i = 1, \ldots, n$ define*

$$P_{bottom}^{[i]} = P^{(1)} + P^{(2)} + \ldots + P^{(i)} \, and,$$
$$P_{up}^{[i]} = P^{(n)} + P^{(n-1)} + \ldots + P^{(n-i+1)},$$

and also define $P_{bottom}^{[0]} = P_{up}^{[0]} = (0,0)$.

The bottom sorted path is the curve that is formed by connecting $P_{bottom}^{[i]}$ and $P_{bottom}^{[i+1]}$ with a line, for $i = 0, \ldots, n-1$. The upper sorted path is the curve that is formed by connecting $P_{up}^{[i]}$ and $P_{up}^{[i+1]}$ with a line, for $i = 0, \ldots, n-1$.

The integral point $P_{bottom}^{[i]}$ on the bottom sorted path corresponds to allocating the i services with the lowest VMR to the first bin and the rest to the second. Similarly, the integral point $P_{up}^{[i]}$ on the upper sorted path corresponds to allocating the i services with the highest VMR to the first bin and the rest to the second. A fractional partition is one that allows splitting a service between several bins. [1] showed that:

Lemma 1. *The polygon confined by the bottom sorted path and the upper sorted path is the convex hull of all possible partitions, i.e. all the fractional and integral points lie within it.*

The second task is analyzing the cost function D. We proved that when the cost function D respects three natural properties (see Sect. 4 below) and when $c_1 \le c_2$ the optimal fractional solution lies on the bottom sorted path and it splits at most one service between the two bins. The implication of that is that it is always better to allocate the low risk services to the smaller capacity bin and the high risk services to the higher capacity bin.

4   Three Cost Functions

Any cost function $D(a,b)$ that has the following three properties falls into the framework of [1]. We give here a brief description of the properties. Full details can be found in [1].

1. <u>Symmetry</u>: $D(a, b) = D(1 - a - \frac{c_2 - c_1}{\mu}, 1 - b)$.
2. <u>Uni-modality in a</u>: For every fixed $b \in [0, 1]$, $D(a, b)$ has a unique minimum on $a \in [0, 1]$, at some point $a = m(b)$. The curve $\{(m(b), b)\}$ is called the *valley*.
3. <u>Central saddle point</u>: D has a unique maximum over the valley at the point $(m(\frac{1}{2}), \frac{1}{2})$.

The above three properties hold for a very large family of cost functions, and in particular for the following three cost functions which are often used in practice. Again, we give here a brief description. Full details and proofs (that these cost functions indeed respect these three properties) can be found in [1].

- **SP-MED** (*Stochastic Placement with Min Expected Deviation*): In SP-MED the cost is the sum of the expected deviations of the bins, i.e. $\sum_{j=1}^{k} \mathbb{E} f_j(X_j)$, where $f_j(x)$ is the deviation function of bin j, i.e., $f_j(x) = x - c_j$ if $x > c_j$ and 0 otherwise, and $\mathbb{E} f_j(X_j)$ is the expected deviation of bin j.
- **SP-MWOP** (*Stochastic Placement with Min Worst Overflow Probability*): In SP-MWOP the cost is the minimal probability p, such that for every bin the probability that the bin overflows is at most p. Namely, if OF_j is the event that bin j overflows, then the cost of a placement is $\max_{j=1}^{k} \Pr[OF_j]$. In the normal case the overflow probability of bin j, denoted by OFP_j, is $OFP_j = 1 - \Phi(\Delta_j)$, where Φ is the cumulative distribution function of the standard normal distribution, $\Delta_j = \frac{c_j - \mu_j}{\sigma_j}$, $\mu_j = \sum_{i \in S_j} \mu^{(i)}$, and $\sigma_j = \sqrt{V_j} = \sqrt{\sum_{i \in S_j} V^{(i)}}$. Thus, $D_{SP-MWOP} = \max_{j=1}^{k} \{1 - \Phi(\Delta_j)\}$.

  We proved in [1] that when $k = 2$, the optimal fractional solution for the SP-MWOP cost function is the unique point on the bottom sorted path, in which $\Delta_1 = \Delta_2$ is satisfied.
- **SP-MOP** (*Stochastic Placement with Minimum Overflow Probability*): In SP-MOP the cost is the probability that *any* bin overflows, i.e. $\Pr[\bigcup_{j=1}^{k} OF_j]$. The total overflow probability is therefore $1 - \prod_{j=1}^{k} (1 - OFP_j)$, where in the normal case, $OFP_j = 1 - \Phi(\Delta_j)$. Thus, $D_{SP-MOP} = 1 - \prod_{j=1}^{k} \Phi(\Delta_j)$.

5   The Double Sorting Framework for More Than Two Data Centers

We now analyze the general $k > 2$ bin case using the results obtained for the two-bin case. We assume that the cost function has the following property:

<u>The Local Optimality condition</u>: In any optimal solution to a k-bin problem, the allocation for any two bins is also optimal. Formally, if $S_1, \ldots, S_k$ is optimal for a k-bin problem, then for any j and j', the partition $S_j, S_{j'}$ is optimal for the two-bin problem defined by the services in $S_j \cup S_{j'}$ and capacities $c_j, c_{j'}$.

Theorem 1. *The cost functions SP-MED, SP-MWOP and SP-MOP respect the local optimality condition.*

Proof. Let $S_1, \ldots, S_k$ be an optimal solution. Suppose there are j and j' for which $S_j, S_{j'}$ are not the optimal solution for the two-bin problem defined by the services in $S_j \cup S_{j'}$ and capacities $c_j, c_{j'}$. Change the allocation of the solution $S_1, \ldots, S_k$ on bins j and j' to an optimal solution for the two bin problem.

- In SP-MWOP, this change improves the worst overflow probability of the two bins while not affecting the overflow probability of any other bin.
- In SP-MED, this change improves the expected total deviation of the two bins while not affecting the expected deviation of any other bin.
- In SP-MOP, this change improves $(1 - OFP_j)(1 - OFP_{j'})$ while not affecting the overflow probability of any other bin.

In total we get a better solution, in contradiction to the optimality of $S_1, \ldots, S_k$.

Definition 2 *(A sorted solution). Assume the bins are sorted by their capacity, $c_1 \leq c_2 \leq \ldots \leq c_k$. A fractional solution is called* sorted *if for every $j < j'$ and every two services i and i' such that service i (resp. i') is allocated to bin j (resp. j') it holds that the VMR of service i is not larger than that of service i'.*

Theorem 2. *If the cost function satisfies the symmetry, unimodality in a, central saddle point and local optimality properties, then the optimal fractional solution is* sorted.

Proof. Assume there is an optimal solution $S_1, \ldots, S_k$ that is *not* sorted. I.e., there exist $j < j'$ and i, i' such that service i (resp. i') is allocated to bin j (resp. j') and the VMR of service i is strictly larger than that of service i'. By [1] the sorted fractional solution for the problem defined by bins j and j' is *strictly better* than the one offered by the solution $S_1, \ldots, S_k$. This means the solution is not optimal for bins j and j' - in contradiction to the local optimality condition.

Next we present an algorithmic framework for the problem:

The double sorting algorithm framework

- Sort the bins by their capacity $c_1 \leq c_2 \leq \ldots \leq c_k$.
- Sort the services by their VMR $\frac{V^{(1)}}{\mu^{(1)}} \leq \frac{V^{(2)}}{\mu^{(2)}} \leq \cdots \leq \frac{V^{(n)}}{\mu^{(n)}}$.
- Use a partitioning algorithm to find $k - 1$ partition points $\ell_0 \triangleq 0 \leq \ell_1 < \ldots \ell_{k-1} \leq \ell_k \triangleq n$ and output $S_1, \ldots, S_k$ accordingly.
- Allocate the services in S_j to bin number j (with capacity c_j).

The algorithmic framework is not complete since it does not specify how to find the partition points on the bottom sorted path. With two bins there is only one partition point, and we can check all possible $n - 1$ integral partition points. With k bins there are $\binom{n}{k-1}$ possible partition points, and checking all partition points may be infeasible. In the sequel, we present three algorithms for finding a good partition and show (using the local optimality condition) that if we know how to solve the two bin case, we can also solve the k-bin case.

6   A Dynamic Programming Algorithm

We now describe a dynamic programming algorithm that finds the best integral solution on the bottom sorted path. It assumes we can solve the $k = 2$ case (for any cost function D). Thus, the algorithm is quite general and may be seen as a general reduction from arbitrary k to the $k = 2$ case. The complexity of the algorithm is polynomial, on the order of about n^3 time.

Finding an integral partition using dynamic programming

For each $1 \leq t \leq \log k$, $1 \leq i < i' \leq n$ and $j = 1, \ldots, k$ keep the two values $Partition(2^t, i, i', j)$ and $D(2^t, i, i', j)$ that are defined as follows:

- Base case $t = 1$: $Partition(2, i, i', j)$ is the best integral partition point for the two bin problem with inputs $X_i, \ldots, X_{i'}$ and capacities c_j, c_{j+1}. $D(2, i, i', j)$ is its corresponding cost.
- Induction step. Suppose we have built the tables for t, we show how to build the tables for $t + 1$. We let $D(2^{t+1}, i, i', j)$ be: $\min \{F(D(2^t, i, i'' - 1, j), D(2^t, i'', i', j + 2^t))\}$, where the minimization is over $i + 2^t \leq i'' \leq i' - (2^t - 1)$ and F depends on the cost function we are dealing with[a]. We let $Partition(2^{t+1}, i, i', j)$ be the partition point that obtains the minimum in the above equation.

The best partition to k bins (assuming k is a power of two) can be recovered from the tables. For example for 4 bins $Partition(4, 1, n, 1)$ returns the middle point ℓ_2 of the partition, and the partition points within each half are obtained by $\ell_1 = Partition(2, 1, \ell_2 - 1, 1)$ and $\ell_3 = Partition(2, \ell_2, n, 3)$.

---

[a] for SP-MED: $F(x, y) = x + y$; for SP-MWOP: $F(x, y) = max\{x, y\}$; for SP-MOP: $F(x, y) = 1 - (1 - x) \cdot (1 - y)$.

It can be seen (by a simple induction) that $Partition(2^t, i, i', j)$ gives the middle point of the best integral solution on the sorted path to the problem of dividing $X_i, \ldots, X_{i'}$ to 2^t bins with capacities $c_j, c_{j+1}, \ldots, c_{j+2^t-1}$. It can also be verified that the running time of the algorithm is $O(n^3 k \log k)$.[1]

---

[1] The best integral solution on the bottom sorted path is not necessarily the optimal integral solution. However, in any reasonable situation, its cost is *close* to the optimal cost, and even to the optimal fractional solution. To see that notice that by [1] the optimal fractional solution is on the bottom sorted path, and if there is no dominant service (see [1] for a rigorous definition) there must be an integral point on the bottom sorted path that is close to the optimal fractional solution, and therefore by continuity (if indeed the cost function is continuous) the cost of the best integral solution on the bottom sorted path is close to the optimal fractional cost. A rigorous analysis for two bins and the cost functions SP-MED and SP-MWOP is presented in [1]. The k bin case is a straightforward extension of these results.

7   The Moving Sticks (MVS) Algorithm for SP-MWOP

The main disadvantage of the dynamic algorithm is that it runs in about n^3 time. For specific cases, such as SP-MWOP cost function, the optimal solution may be found much faster.

We proved in [1] that when $k = 2$, the optimal fractional solution for SP-MWOP cost function is the unique point on the bottom sorted path, in which the equality $\Delta_1 = \Delta_2$ is satisfied. Hence, by the local optimality condition we can say that the optimal fractional solution of SP-MWOP for k-bins is obtained on a partition where $\Delta_1 = \Delta_2 = \ldots = \Delta_k$.

The dynamic algorithm finds an integral solution. In the scope of the moving sticks (MVS) algorithm, it is more convenient to find a fractional solution and work with fractional partition point values. Suppose we have n services sorted by their VMR, and service i has mean $\mu^{(i)}$. The first m services correspond to the fractional point $f = \frac{\sum_{i=1}^{m} \mu^{(i)}}{\mu} \in [0,1]$. Conversely, a point $f \in [0,1]$ with $\frac{\sum_{i=1}^{m} \mu^{(i)}}{\mu} \leq f < \frac{\sum_{i=1}^{m+1} \mu^{(i)}}{\mu}$ corresponds to taking the first m services, and the appropriate part from service $m + 1$. Similarly, a segment $[\alpha_1, \alpha_2] \subseteq [0,1]$ corresponds to a consecutive set of services, with up to two incomplete services (from the beginning and end of the segment). A fractional partition is $0 = f_0 < f_1 < \ldots < f_{k-1} < f_k = 1$.

Following is an informal high level description of the algorithm: At each time step of the algorithm we have $k - 1$ sticks positioned between the n services (sorted in increasing order by their VMR) and thus separating them to k bins, where the j'th segment is allocated to the j'th bin. At the beginning, we fill the last bins to full capacity as much as we can. After initialization, there are some completely full bins at the end of the bin list, at most one bin which is partially full, and zero or more empty bins at the beginning of the list. It is easy to see that in the initial state each partition point is left to its position in the fractional optimal solution. Also, at the first step $\Delta_1 \geq \Delta_2 \geq \ldots \geq \Delta_k$. Then, at each time step, we make a local improvement by fractionally moving one of the sticks right, provided that this movement does not break the invariance that $\Delta_1 \geq \Delta_2 \geq \ldots \geq \Delta_k$. We will see that this guarantees that again each partition point is left to its position in the fractional optimal solution. The process then continues until no local improvement is possible. Using the fact that the SP-MWOP fractional optimal partition for k bins is achieved when $\Delta_1 = \Delta_2 = \ldots = \Delta_k$, we prove that the process converges to the optimal fractional solution on the bottom sorted path.

The fractional step size τ, is the fraction of the total mean, μ, moved from one bin to its predecessor in one step of the algorithm. τ determines the algorithm running time as well as its accuracy. Since there are $k - 1$ sticks, and each stick can move at most $\frac{1}{\tau}$ times, the algorithm has at most $\frac{k-1}{\tau}$ steps before it stops. Moreover, the smaller τ is the better the accuracy is (and, correspondingly, the running time gets larger). In Theorem 3 we show that the algorithm error bound linearly depends on k and τ.

The moving sticks algorithm for SP-MWOP: Initialization

- **Initialization** (at time $t = 0$): /* completely fill up bins */
  - Let $\tilde{j}$ be the smallest bin index which satisfies $c_{\tilde{j}+1} + c_{\tilde{j}+2} + \ldots + c_k < \mu$.
  - $\forall j = 0, 1, \ldots, k$, let

$$f_j^{(0)} = \begin{cases} 0 & \text{If } j < \tilde{j} \\ 1 - \frac{c_{j+1}+c_{j+2}+\ldots+c_k}{\mu} & \text{Else} \end{cases}$$

  - Let $I^{(0)} = \{2, \ldots, k\}$.[a]

[a] We could have taken $I^{(0)} = \{\tilde{j}, \tilde{j}+1\}$, but this does not improve the asymptotic behavior of the algorithm, and we choose the simpler rule.

A detailed description of the moving sticks algorithm for the SP-MWOP is given below. We later prove its correctness. For the description we let $\Delta_j^{[\alpha_1, \alpha_2]}$ denote the value of Δ_j, when bin j is (fractionally) allocated the items corresponding to the segment $[\alpha_1, \alpha_2] \subseteq [0, 1]$. We also let $I^{(t)}$ denote the set of bin indices that can potentially move $\tau \cdot \mu$ to their predecessor without violating the invariance.

The moving sticks algorithm for SP-MWOP: Main loop

- **Main loop:** Start with $t = 0$. Repeat until $I = \emptyset$:
  - Pick an index $j \in I^{(t)}$.
  - Calculate:
    - $nextΔ_{j-1} = \Delta_{j-1}^{[f_{j-2}^{(t)}, f_{j-1}^{(t)}+\tau]}$,
    - $nextΔ_j = \Delta_j^{[f_{j-1}^{(t)}+\tau, f_j^{(t)}]}$.

  - If $(nextΔ_j > nextΔ_{j-1})$ /* we break the invariance */
    - Remove index j from I. /* do not move any stick */

  - Else,
    - Set $f_{j-1}^{(t+1)} = f_{j-1}^{(t)} + \tau$ and keep all other values unchanged, i.e., $f_{j'}^{(t+1)} = f_{j'}^{(t)}, \forall j' \neq j - 1$.
    - Let

$$I^{(t+1)} = \begin{cases} I^{(t)} \cup \{3\} & \text{If } j = 2 \\ I^{(t)} \cup \{k-1\} & \text{If } j = k \\ I^{(t)} \cup \{j-1, j+1\} & \text{Else} \end{cases}$$

  - Increase t.
- **Stopping rule:** When $I = \emptyset$ stop and output $f_1^{(t)}, \ldots, f_{k-1}^{(t)}$.

We now prove the correctness of the moving sticks algorithm for the SP-MWOP problem.

Theorem 3. *Let* $0 = f_0^* \leq f_1^* \leq f_2^* \leq \ldots \leq f_{k-1}^* \leq f_k^* = 1$ *be the optimal fractional partition. The following is true for the moving sticks algorithm, when considering the SP-MWOP problem:*

1. *At time $t = 0$, no partition point is positioned to the right to its optimal position, i.e.,*
   $$f_j^{(0)} \leq f_j^*, \forall j \in [0, k]$$
2. *At any time step $t \geq 0$, each separator may move only rightward, i.e.,*
   $$f_j^{(t+1)} \geq f_j^{(t)}, \forall j \in [1, k-1]$$
3. *For $t \geq 0$, define $\Delta_j^{(t)} = \Delta_j^{[f_{j-1}^{(t)}, f_j^{(t)}]}$. At any time step $t \geq 0$,*

$$\Delta_1^{(t)} \geq \Delta_2^{(t)} \geq \ldots \geq \Delta_k^{(t)}. \tag{1}$$

4. *If $f_j^{(t)} \leq f_j^* \ \forall j \in [1, k-1]$, then also $f_j^{(t+1)} \leq f_j^* \ \forall j \in [1, k-1]$.*
5. *Stopping state: when we stop, we cannot move any stick right by τ without breaking the algorithm invariance (1).*
6. *Error bounding: Let VMR_{max} be the maximum VMR value of the input, i.e. $VMR_{max} = \frac{V^{(n)}}{\mu^{(n)}}$, and let b_{min} be the minimum variance portion a bin gets. The difference between the cost of the fractional solution found by the moving sticks algorithm and the optimal fractional solution is at most $(k-1)\epsilon$, where $\epsilon = \frac{c}{\sigma} \cdot \frac{\sqrt{2}(1+VMR_{max})}{b_{min}\sqrt{b_{min}}} \cdot \tau$.*

For the proof of Theorem 3 we need two technical lemmas.

Lemma 2 *(monotonicity of an optimal partition point). Let $OPT_j^{[\alpha_1, \alpha_2]}$ denote the SP-MWOP fractional partition point which optimally allocates the items that belong to the segment $[\alpha_1, \alpha_2]$ to the two bins j and $j + 1$. Let $1 \leq j \leq k - 1$. Then:*

1. *For all $\alpha_1 \leq \alpha_2 < \alpha_3 \leq 1$, $OPT_j^{[\alpha_1, \alpha_3]} \leq OPT_j^{[\alpha_2, \alpha_3]}$.*
2. *For all $\alpha_1 < \alpha_2 \leq \alpha_3 \leq 1$, $OPT_j^{[\alpha_1, \alpha_2]} \leq OPT_j^{[\alpha_1, \alpha_3]}$.*

Proof. We prove the first item, the proof of the second is similar. Recall that an SP-MWOP fractional optimal partition for two bins is achieved at the point in which $\Delta_1 = \Delta_2$, i.e., $\Delta_j^{[\alpha_2, OPT_j^{[\alpha_2, \alpha_3]}]} = \Delta_{j+1}^{[OPT_j^{[\alpha_2, \alpha_3]}, \alpha_3]}$. If we add the items in the segment $[\alpha_1, \alpha_2]$ to bin j, we decrease its spare capacity and get that $\Delta_j^{[\alpha_1, OPT_j^{[\alpha_2, \alpha_3]}]} < \Delta_j^{[\alpha_2, OPT_j^{[\alpha_2, \alpha_3]}]}$ and therefore, $\Delta_j^{[\alpha_1, OPT_j^{[\alpha_2, \alpha_3]}]} < \Delta_{j+1}^{[OPT_j^{[\alpha_2, \alpha_3]}, \alpha_3]}$. In order to reach the optimal partition in the segment $[\alpha_1, \alpha_3]$, we need to decrease Δ_{j+1} and increase Δ_j till we get equality, i.e., we need to move items from bin j to bin $j + 1$, by decreasing the partition point between them. Hence, we get that $OPT_j^{[\alpha_1, \alpha_3]} \leq OPT_j^{[\alpha_2, \alpha_3]}$.

Lemma 3 *(continuity of Δ).* *Let VMR_{max}, b_{min} as before. Let $P_1 = (\alpha_1, \beta_1)$,*
$P'_1 = (\alpha'_1, \beta'_1)$, $P_2 = (\alpha_2, \beta_2)$ and $P'_2 = (\alpha'_2, \beta'_2)$, be four points on the bottom
sorted path, such that $0 \le \alpha_1 < \alpha'_1 < \alpha_2 < \alpha'_2 \le 1$ (and therefore, $0 \le \beta_1 \le$
$\beta'_1 \le \beta_2 \le \beta'_2 \le 1$), and also $\alpha'_1 - \alpha_1 = \tau$ and $\alpha'_2 - \alpha_2 = \tau$. Then, for all
$1 \le j \le k$: $\Delta_j^{[\alpha_1,\alpha_2]} - \Delta_j^{[\alpha_1,\alpha'_2]} \le \epsilon$, and, $\Delta_j^{[\alpha_1,\alpha'_2]} - \Delta_j^{[\alpha'_1,\alpha'_2]} \le \epsilon$, where $\epsilon =$
$\frac{c}{\sigma} \cdot \frac{\sqrt{2}(1+VMR_{max})}{b_{min}\sqrt{b_{min}}} \cdot \tau$.

Proof. We recall a notation we have used in Sect. 3: $\Delta_j(a,b) = \frac{c_j - a\mu}{\sigma\sqrt{b}}$, where the
demand to bin j is normally distributed with mean $a\mu$ and variance bV. In this
notation, and using the mean value theorem:

$$\Delta_j^{[\alpha_1,\alpha_2]} - \Delta_j^{[\alpha_1,\alpha'_2]} = \Delta_j(\alpha_2 - \alpha_1, \beta_2 - \beta_1) - \Delta_j(\alpha'_2 - \alpha_1, \beta'_2 - \beta_1) \le \quad (2)$$
$$|\nabla(\Delta_j(a,b))| \cdot |(\alpha_2 - \alpha_1, \beta_2 - \beta_1) - (\alpha'_2 - \alpha_1, \beta'_2 - \beta_1)|,$$

where $\alpha_2 - \alpha_1 \le a \le \alpha'_2 - \alpha_1$ and $\beta_2 - \beta_1 \le b \le \beta'_2 - \beta_1$.

It is easy to bound the gradient of $\Delta(a,b)$:

$$\left|\frac{\partial \Delta_j(a,b)}{\partial a}\right| = \frac{\mu}{\sigma} \cdot \frac{1}{\sqrt{b}} \le \frac{c}{\sigma} \cdot \frac{1}{\sqrt{b}}$$
$$\left|\frac{\partial \Delta_j(a,b)}{\partial b}\right| = \frac{1}{2} \cdot \frac{c_j - a\mu}{\sigma} \cdot \frac{1}{b\sqrt{b}} \le \frac{c}{\sigma} \cdot \frac{1}{b\sqrt{b}}$$
$$|\nabla(\Delta_j(a,b))| \le \frac{c}{\sigma} \cdot \frac{1}{\sqrt{b}} \cdot \sqrt{1 + \frac{1}{b^2}} = \frac{c}{\sigma} \cdot \frac{\sqrt{1+b^2}}{b\sqrt{b}} \le \frac{c}{\sigma} \cdot \frac{\sqrt{2}}{b\sqrt{b}}.$$

Since $b \ge b_{min}$ we get that: $|\nabla(\Delta_j(a,b))| \le \frac{c}{\sigma} \cdot \frac{\sqrt{2}}{b_{min}\sqrt{b_{min}}}$.

Finally, $|(\alpha_2 - \alpha_1, \beta_2 - \beta_1) - (\alpha'_2 - \alpha_1, \beta'_2 - \beta_1)| \le |\alpha'_2 - \alpha_2| + |\beta'_2 - \beta_2|$. We
know that $|\alpha'_2 - \alpha_2| \le \tau$. Also, $\frac{\beta'_2 - \beta_2}{\alpha'_2 - \alpha_2} \le VMR_{max}$. Therefore, we can say that
$|(a',b') - (a'',b'')| \le \tau + \tau \cdot VMR_{max} = \tau(1 + VMR_{max})$. When we plug this in
Eq. (2) we get the lemma.

We are now ready to prove Theorem 3:

Proof. We go over the items one by one:

Item (1): For every $0 < j \le \tilde{j}$ the algorithm sets $f_{j-1}^{(0)} = 0$, which is obviously
less than or equal to the optimal value f_{j-1}^*. Also, for every $\tilde{j} < j \le k$,
the algorithm initializes the j'th bin to be completely full. In all partitions
(including the optimal fractional partition) no bin is allocated more than its
capacity. Hence, again, $f_{j-1}^{(0)} \le f_{j-1}^*$. Finally, $f_k^{(0)} = f_k^* = 1$.

Item (2): At each time step, the algorithm increases one separator value by τ
(i.e., moves this separator right) and leaves all others unchanged. Therefore,
$f_j^{(t+1)} \ge f_j^{(t)}, \forall j \in [1, k-1]$.

Item (3): After initialization, $\Delta_j^{(0)} = 0$ for every completely full bin $\tilde{j} < j \le k$,
$0 < \Delta_j^{(0)} < \infty$ for the only partially full bin $j = \tilde{j}$, and $\Delta_j^{(0)} = \infty$ for every
empty bin $j < \tilde{j}$. Therefore, after initialization the inequality $\Delta_1^{(0)} \ge \Delta_2^{(0)} \ge$
$\ldots \ge \Delta_k^{(0)}$ holds.

Let us assume that at the end of time $t > 0$ the inequality $\Delta_1^{(t)} \geq \Delta_2^{(t)} \geq \ldots \geq \Delta_k^{(t)}$ holds. At time $t+1$, the algorithm picks bin j' and checks whether to move stick $j'-1$ and update $f_{j'-1}$. Since all other sticks are not updated, $\Delta_j^{(t+1)} = \Delta_j^{(t)}$, $\forall j \notin \{j'-1, j'\}$. In addition, the stick is moved only if the inequality $\Delta_{j'-1}^{(t+1)} \geq \Delta_{j'}^{(t+1)}$ stays true after the move. And since the move decreases $\Delta_{j'}$ (i.e. $\Delta_{j'}^{(t+1)} \leq \Delta_{j'}^{(t)}$) and increases $\Delta_{j'-1}$ (i.e. $\Delta_{j'-1}^{(t+1)} \geq \Delta_{j'-1}^{(t)}$), the inequality $\Delta_1^{(t+1)} \geq \Delta_2^{(t+1)} \geq \ldots \geq \Delta_k^{(t+1)}$ is also true.

Item (4): At each time step t, the algorithm picks bin $j' \in \{2, \ldots, k\}$ and moves a fraction of an item from bin j' to bin $j'-1$, as long as the invariance $\Delta_{j'-1} \geq \Delta_{j'}$ stays true after the move. All other bins are kept untouched and therefore $f_j^{(t+1)} = f_j^{(t)} \leq f_j^*$ $\forall j \neq j'-1$.

Now, let us consider bin $j'-1$ and bin j'. The list of items packed in both bins at time t is the list of items which corresponds to the segment $[f_{j'-2}^{(t)}, f_{j'}^{(t)}]$ and the algorithm, by moving fraction of items from bin j' to bin $j'-1$, increases $f_{j'-1}^{(t)}$ provided that the inequality $\Delta_{j'-1} \geq \Delta_{j'}$ stays valid. Since the optimal fractional solution is achieved when $\Delta_{j'-1} = \Delta_{j'}$, we get that

$$f_{j'-1}^{(t+1)} \leq OPT_{j'-1}^{[f_{j'-2}^{(t)}, f_{j'}^{(t)}]}. \tag{3}$$

In the optimal fractional partition the list of items packed in bins $j'-1$ and j' correspond to the segment $[f_{j'-2}^*, f_{j'}^*]$. We are given that $f_j^{(t)} \leq f_j^*$ $\forall t$ and $j \in \{1, \ldots, k-1\}$, and in particular $f_{j'-2}^{(t)} \leq f_{j'-2}^*$ and $f_{j'}^{(t)} \leq f_{j'}^*$. By applying Lemma 2 twice (once with $\alpha = f_{j'-2}^{(t)}$, $\beta_1 = f_{j'}^{(t)}$ and $\beta_2 = f_{j'}^*$, and once with $\alpha_1 = f_{j'-2}^{(t)}$, $\alpha_2 = f_{j'-2}^*$ and $\beta = f_{j'}^*$) we conclude that

$$OPT_{j'-1}^{[f_{j'-2}^{(t)}, f_{j'}^{(t)}]} \leq f_{j'-1}^*. \tag{4}$$

Equations (3) and (4) together imply that $f_{j'-1}^{(t+1)} \leq f_{j'-1}^*$.

Item (5): We will show that at any time t, bin indices which are not included in $I^{(t)}$ cannot move $\tau \cdot \mu$ to their preceding bin without breaking the algorithm invariance. At the beginning, $I^{(t=0)} = \{2, \ldots, k\}$, so there is no bin outside $I^{(t=0)}$. Assume the property is true for time t, and let us prove correctness for time $t+1$. At time t, the algorithm picks an index $j \in I$, and a fractional movement of items from bin j to bin $j-1$ is examined. If the movement breaks the invariance, no movement is done and j is justifiably removed from I, i.e. $I^{(t+1)} = I^{(t)} \setminus \{j\}$. If the movement does not break the invariance, then the move is executed ($f_{j'-1}^{(t+1)}$ is updated) and the bins that are affected (i.e., bin indices $j-1$ and $j+1$) are added to I. Hence the required property is preserved.

Hence, when we stop, $I^{(t)} = \emptyset$, i.e., a move of $\tau \cdot \mu$ from any bin j to bin $j-1$ will necessarily break the invariance, as we had to prove.

Item (6): Let $0 = f_0^* \le f_1^* \le \ldots \le f_k^* = 1$ be the partition points of the fractional optimal solution, and $0 = f_0 \le f_1 \le \ldots \le f_k = 1$ be the partition points found by the moving sticks algorithm. Similarly, Δ_j^* (resp. Δ_j) is the j'th bin Δ value in the partition f^* (resp. f). In the partition f^* we have $\Delta_1^* = \ldots = \Delta_k^*$. We now show that when the moving sticks algorithm stops all the Δ_j values are close to each other. Namely,

$\quad$ *Claim.* For every $1 \le j \le k-1$: $\Delta_j \le \Delta_{j+1} + 2\epsilon$.

Proof. Let f_j' be the optimum partition point for the two bins problem j and $j+1$, with the item list in segment $[f_{j-1}, f_{j+1}]$. From the algorithm invariance we know that $\Delta_j^{[f_{j-1}, f_j]} \ge \Delta_{j+1}^{[f_j, f_{j+1}]}$. On the other hand, $\Delta_j^{[f_{j-1}, f_j']} = \Delta_{j+1}^{[f_j', f_{j+1}]}$, since f_j' is an optimum partition point. Therefore, we get that $f_j \le f_j'$. Moreover, by item (3) $f_j' - f_j < \tau$, and by Lemma 3, $\Delta_j^{[f_{j-1}, f_j]} - \Delta_j^{[f_{j-1}, f_j']} \le \epsilon$ and $\Delta_{j+1}^{[f_j', f_{j+1}]} - \Delta_{j+1}^{[f_j, f_{j+1}]} \le \epsilon$. Therefore,

$$\Delta_j = \Delta_j^{[f_{j-1}, f_j]} \le \Delta_j^{[f_{j-1}, f_j']} + \epsilon = \Delta_{j+1}^{[f_j', f_{j+1}]} + \epsilon \le \Delta_{j+1}^{[f_j, f_{j+1}]} + 2\epsilon = \Delta_{j+1} + 2\epsilon.$$

$\quad$ The claim implies that $\Delta_k \ge \Delta_{k-1} - 2\epsilon \ge \Delta_{k-2} - 4\epsilon \ge \ldots \ge \Delta_1 - 2(k-1)\epsilon$. However, from Theorem 3 we know that $f_{k-1} \le f_{k-1}^*$ and we also know that $f_k = f_k^*$. This means that $[f_{k-1}^*, f_k^*] \subseteq [f_{k-1}, f_k]$, and hence $\Delta_k \le \Delta_k^*$. Using similar arguments, we can say that $\Delta_1 \ge \Delta_1^*$. Together, $\Delta_k^* \ge \Delta_k \ge \Delta_1 - 2(k-1)\epsilon \ge \Delta_1^* - 2(k-1)\epsilon = \Delta_k^* - 2(k-1)\epsilon$.

$\quad$ The algorithm invariance is $\Delta_1 \ge \Delta_2 \ge \ldots \ge \Delta_k$, and this is also true when the algorithm ends. The cost of the fractional solution found by the moving sticks algorithm is therefore $max_{j=1}^k \{1 - \Phi[\Delta_j]\} = 1 - \Phi[\Delta_k]$. Also, the cost of the fractional optimal solution is $1 - \Phi[\Delta_k^*]$. The error is therefore $\Phi[\Delta_k^*] - \Phi[\Delta_k]$, where we know that $|\Delta_k - \Delta_k^*| \le 2(k-1)\epsilon$. By the mean value theorem the difference is at most $\phi[\Delta_\xi] \cdot 2(k-1)\epsilon \le (k-1)\epsilon$, where $\Delta_k^* - 2(k-1)\epsilon \le \Delta_\xi \le \Delta_k^*$.

8   The Generalized Moving Sticks (GMVS) Algorithm

In the previous section we introduced the moving sticks algorithm for the SP-MWOP cost function and gave a correctness proof as well as error analysis. In this section we present a more general form of the moving sticks algorithm for other cost functions, such as SP-MED and SP-MOP, without a correctness proof and without error bounding. Simulations show that the general moving stick (GMVS) algorithm performs extremely well for all three cost functions we consider in this paper.

$\quad$ A key observation for generalizing the moving sticks algorithm for other cost functions is that the cost on the bottom sorted path, when expressed as a function of a,[2] is unimodal in the two bin case when $\Delta_1 \ge 0$ and $\Delta_2 \ge 0$, i.e., decreasing till the optimal point and increasing from there on. Therefore, we can leave the initialization as is, and change only part of the main loop, such that

---

[2] Where a is the portion of the total mean allocated to the first bin, i.e., $a = \frac{\mu_1}{\mu}$.

at each time step, there is a fractional move of one of the sticks right, provided that this movement decreases the two-bin cost function of the two bins left and right of it. The advantage of this generalized version is that it is applicable to many cost functions, such as SP-MED, SP-MWOP and SP-MOP.

Following is a detailed description of the generalized main loop. For the description, we let $D_j^{[\alpha_1,\alpha_2,\alpha_3]}$ denote the cost of the two bins j and $j+1$, where bin j is (fractionally) allocated the items corresponding to the segment $[\alpha_1,\alpha_2] \subseteq [0,1]$ and bin $j+1$ is (fractionally) allocated the items corresponding to the segment $[\alpha_2,\alpha_3] \subseteq [0,1]$.

The generalized moving sticks algorithm: Main loop

– **Main loop:** Start with $t = 0$. Repeat until $I = \emptyset$:
  - Pick an index $j \in I^{(t)}$.
  - Calculate:
    - $currentD = D_{j-1}^{[f_{j-2}^{(t)},f_{j-1}^{(t)},f_j^{(t)}]}$.
    - $nextD = D_{j-1}^{[f_{j-2}^{(t)},f_{j-1}^{(t)}+\tau,f_j^{(t)}]}$,
  - If $(nextD > currentD)$ /* two-bin cost is increased */
    - Remove index j from I. /* do not move any stick */
  - Else,
    - Set $f_{j-1}^{(t+1)} = f_{j-1}^{(t)} + \tau$ and keep all other values unchanged, i.e., $f_{j'}^{(t+1)} = f_{j'}^{(t)}, \forall j' \neq j-1$.
    - Let

$$I^{(t+1)} = \begin{cases} I^{(t)} \cup \{3\} & \text{If } j = 2 \\ I^{(t)} \cup \{k-1\} & \text{If } j = k \\ I^{(t)} \cup \{j-1, j+1\} & \text{Else} \end{cases}$$

  - Increase t.
– **Stopping rule:** When $I = \emptyset$ stop and output $f_1^{(t)}, \ldots, f_{k-1}^{(t)}$.

---

9 Conclusions

We present an analytical scheme for stochastic placement algorithms, using the stochastic behavior of the demand. We generalize the results of previous work, which considered only two data centers, and develop efficient, almost optimal algorithms that work for a family of target cost functions and any number of data centers. In particular, we solve SP-MED (that minimizes the expected deviation), SP-MOP (that minimizes the probability of overflow) and SP-MWOP (that guarantees that for every bin the probability it overflows is small). We believe the framework is applicable for many other natural cost functions. As in the case of only two data centers, the results in this paper hold for any large enough collection of independent services of whatever distribution.

We show that if we know how to solve the two bin case, we can also solve the $k > 2$ bin case and present three algorithms for finding a good placement. The first algorithm is a dynamic programming algorithm, which finds the best integral point on the bottom sorted path and runs in about n^3 time. The second, is the moving sticks algorithm for the SP-MWOP cost function. We prove that it converges to the optimal solution in almost linear time. The last algorithm is a general version of the moving sticks algorithm, which is applicable for other cost functions and also runs in almost linear time. However, we do not prove its correctness.

Our simulation results indicate that our algorithms provide considerable gain compared with commonly used naive solutions.

A   Simulation Results

In this section we present our simulation results for k bins. In all simulations we used 4 bins with bin capacity ratio of $c_{i+1} = 2c_i$. For each cost function, we compared the generalized moving sticks (GMVS) algorithm with two algorithms we call BS (Balanced Spares) and BL (Balanced Load), as in [1]. The BS algorithm goes through the list, item by item, and allocates each item to the bin which has more available space. In this way, the spare capacity is balanced. On the other hand, the BL algorithm goes through the list, item by item, and allocates each item to the bin which is less loaded, i.e., the bin with higher $\frac{\text{available space}}{\text{bin capacity}}$ value. In this way, the bin load is balanced. The BL and BS algorithms are natural benchmarks and also much better than other naive solutions like first-fit and first-fit decreasing.

We show two different results for the GMVS algorithm: *Fractional GMVS* and *Integral GMVS*. The fractional GMVS algorithm uses the fractional partition output of the generalized moving sticks algorithm, as described in Sect. 8. The integral GMVS algorithm takes this fractional partition and converts it to an integral partition by moving each stick left to next integral point.[3]

In the case of SP-MWOP cost function, we also added results for the *Fractional MVS* and *Integral MVS* algorithms. As before, the fractional MVS algorithm uses the fractional partition output of the moving sticks algorithm, as described in Sect. 7, and the integral MVS algorithm takes this fractional partition and converts it to an integral partition by moving each stick left to next integral point.

We implemented one straight forward improvement for the runtime of both MVS and GMVS algorithms: before examining a fractional movement of an item, we examine the movement of the whole item at once. In this way, if the entire item can be moved from one bin to the other at a given stage, we move it in one step, instead of fractionally moving it in several consecutive steps. This improvement does not change the results of the algorithms, but it improves the running time. Note that we didn't check the extent of this improvement, since we only implemented this version of the algorithms.

A.1   Results for Synthetic Normally Distributed Data

The synthetic data for this part was generated exactly as explained in [1] for the 2 bin case, except that we generated only 400 elements instead of 500 elements

---

[3] Note that we could probably improve results by moving each stick to the closest integral point, which is either left or right of it. However, we think that this improvement is minor when n gets larger.

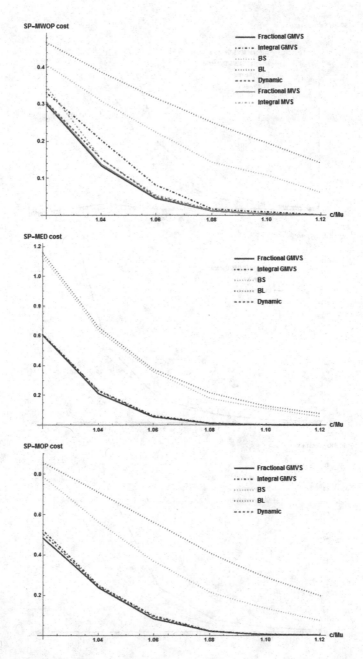

Fig. 1. Average cost of the Fractional and Integral GMVS and MVS algorithms, and the BS and BL algorithms for SP-MWOP, SP-MED and SP-MOP with four bins and synthetic normally distributed data. $n = 400$. $\tau = 0.001$. The x axis measures $\frac{c}{\mu}$.

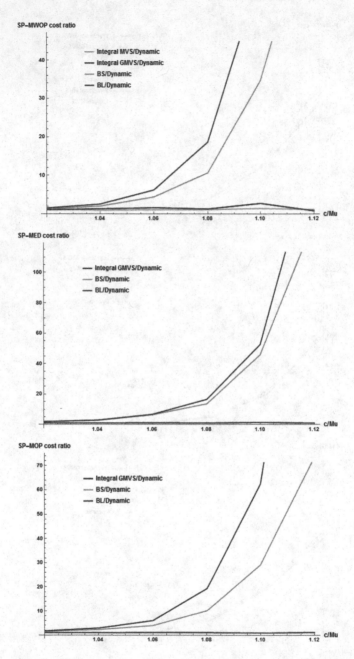

Fig. 2. The BL and BS algorithm costs divided by the dynamic algorithm cost for SP-MWOP, SP-MED and SP-MOP with four bins and synthetic normally distributed data. $n = 400$. $\tau = 0.001$. The x axis measures $\frac{c}{\mu}$.

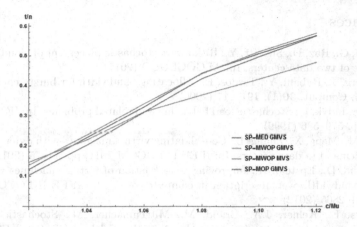

Fig. 3. Average main loop number of iterations (t) divided by number of services (n) in the GMVS and MVS algorithms for the various cost functions with four bins and synthetic normally distributed data. $n = 400$. $\tau = 0.001$. The x axis measures $\frac{c}{\mu}$.

because executing the dynamic algorithm (whose complexity is $O(n^3)$) takes too long on $n = 500$.

Figure 1 shows that for SP-MED and SP-MOP cost functions the Integral GMVS is very close to the Fractional GMVS, as expected. However, for SP-MWOP the integral GMVS and integral MVS are a bit higher that their fractional version. We believe that this gap will be reduced as the number of services is increased (i.e. as n is increased). We also see that for all three cost functions the dynamic algorithm cost is higher that the fractional version of the GMVS and MVS algorithm costs, yet very close to them. Also, the BL and BS algorithm results for the three cost functions are much higher than those of the dynamic algorithm. The integral GMVS, integral MVS, BL and BS algorithm costs divided by the dynamic algorithm cost for the three cost functions is shown in Fig. 2.

To get a feeling of the main loop runtime complexity in both GMVS and MVS algorithms, we calculated the number of main loop iterations (t) in each algorithm run and divided it by the number of services (n). The average results are shown in Fig. 3. We can see that in all cases the number of iterations dose not surpass $0.6n$. We also see that the number of iterations increases as the spare capacity increases. When the spare capacity increases, the bins are less full. Therefore, the final bin of each service is more likely to be different than (and more distant from) it's initial bin, and hence more service movements are required, i.e., more main loop iterations are required. In any case, the process always terminates quickly.

References

1. Shabtai, G., Raz, D., Shavitt, Y.: Risk aware stochastic placement of cloud services: the case of two data centers. In: ALGOCLOUD (2017)
2. Kleinberg, J., Rabani, Y., Tardos, É.: Allocating bandwidth for bursty connections. SIAM J. Comput. **30**(1), 191–217 (2000)
3. Goel, A., Indyk, P.: Stochastic load balancing and related problems. In: IEEE FOCS 1999, pp. 579–586 (1999)
4. Wang, M., Meng, X., Zhang, L.: Consolidating virtual machines with dynamic bandwidth demand in data centers. In: IEEE INFOCOM 2011, pp. 71–75 (2011)
5. Breitgand, D., Epstein, A.: Improving consolidation of virtual machines with risk-aware bandwidth oversubscription in compute clouds. In: IEEE INFOCOM 2012, pp. 2861–2865 (2012)
6. Nikolova, E., Kelner, J.A., Brand, M., Mitzenmacher, M.: Stochastic shortest paths via quasi-convex maximization. In: Azar, Y., Erlebach, T. (eds.) ESA 2006. LNCS, vol. 4168, pp. 552–563. Springer, Heidelberg (2006). https://doi.org/10.1007/11841036_50
7. Nikolova, E.: Approximation algorithms for offline risk-averse combinatorial optimization. In: Approximation, Randomization, and Combinatorial Optimization. Algorithms and Techniques, pp. 338–351 (2010)

Automatic Scaling of Resources in a Storm Topology

Evangelos Gkolemis, Katerina Doka$^{(\boxtimes)}$, and Nectarios Koziris

Computing Systems Laboratory, National Technical University of Athens,
Athens, Greece
{vgol,katerina,nkoziris}@cslab.ece.ntua.gr

Abstract. In the Big Data era, the batch processing of large volumes of data is simply not enough - data needs to be processed fast to support continuous reactions to changing conditions in real-time. Distributed stream processing systems have emerged as platforms of choice for applications that rely on real-time analytics, with Apache Storm [2] being one of the most prevalent representatives. Whether deployed on physical or virtual infrastructures, distributed stream processing systems are expected to make the most out of the available resources, i.e., achieve the highest throughput or lowest latency with the minimum resource utilisation. However, for Storm - as for most such systems - this is a cumbersome trial-and-error procedure, tied to the specific workload that needs to be processed and requiring manual tweaking of resource-related topology parameters. To this end, we propose ARiSTO, a system that automatically decides on the appropriate amount of resources to be provisioned for each node of the Storm workflow topology based on user-defined performance and cost constraints. ARiSTO employs two mechanisms: a *static*, model-based one, used at bootstrap time to predict the resource-related parameters that better fit the user needs and a *dynamic*, rule-based one that elastically auto-scales the allocated resources in order to maintain the desired performance even under changes in load. The experimental evaluation of our prototype proves the ability of ARiSto to efficiently decide on the resource-related configuration parameters, maintaining the desired throughput at all times.

1 Introduction

In the Big Data era, data is being produced not only in large volumes, but also at an astounding rate. Now more than ever, organizations and companies worldwide heavily rely on the processing of the enormous amounts of data that continuously stream into their businesses to extract significant value out of them: identify new risks and opportunities, take educated decisions based on real-time facts, render their operations faster and more cost efficient and keep customers satisfied [7]. The traditional batch processing model is simply not enough any more, since it fails to cover one of Big Data's most important Vs, that of Velocity. Indeed, data needs to be processed fast, to support timely reaction to changing conditions in

© Springer International Publishing AG, part of Springer Nature 2018
D. Alistarh et al. (Eds.): ALGOCLOUD 2017, LNCS 10739, pp. 157–169, 2018.
https://doi.org/10.1007/978-3-319-74875-7_10

real time. This is of paramount importance in sectors such as trading, fraud detection, system monitoring, healthcare and many others [6,11].

A plethora of distributed stream processing engines have emerged as a remedy to the inability of batch processing systems to provide real-time, interactive responses [1,2,9,17]. Such systems are designed to analyze data in motion, as they stream through the server, contrarily to the traditional batch processing model where data are first stored and then subsequently processed by queries.

Distributed stream processing systems are deployed either on bare-metal clusters, or, most often, over Cloud infrastructures which provide virtual resources in a pay-as-you-go manner. The Cloud Computing model offers the ability to elastically allocate resources, i.e., expand and contract them to meet application needs while keeping the resource budget to a minimum. Whether deployed on physical or virtual infrastructures, distributed stream processing systems are expected to make the most out of the available resources, i.e., achieve the highest throughput or lowest latency with the minimum resource utilisation.

Resource provisioning is one of the most challenging tasks for streaming applications, as it is closely related to the rate of data arrival, which can not be controlled since data are generated by external sources. Over-provisioning of resources will unnecessarily increase resource utilization, hence the cost of running the application. Contrarily, under-provisioning may lead to the inability of the application to keep pace with the velocity of the incoming data stream or comply to the throughput or latency target desired by the user.

However, distributed stream processing engines lack mechanisms to assist the application provider to carefully and correctly provision the required resources, let alone provide on-the-fly adaptation to changing workload conditions [4]. Setting the resource-related parameters for such systems is a cumbersome trial-and-error procedure, tied to the specific workload that needs to be processed and requiring manual tweaking. Moreover, sudden changes in the initial setup, e.g., in the data arrival rate or underlying hardware performance, can not be accommodated automatically, but require manual scaling of the allocated resources.

To this end, we propose methods for the automatic provisioning and online scaling of distributed stream processing resources and design a system that implements them on top of Apache Storm, one of the most prevalent distributed streaming engines. Our system, called *ARiSto* (*Auto-Scaling Resources in Storm*), automatically and dynamically decides on the appropriate amount of resources to be provisioned for each node of the Storm workflow topology based on user-defined performance and cost/budget constraints. ARiSto employs two mechanisms:

- A *static* mechanism that is used one-off, when first deploying the topology. This mechanism relies on models of the cost and performance characteristics of the required tasks of the topology graph to decide on the (near-)optimal allocation of resources to each one of them.
- A *dynamic* mechanism that is used on-line, as the application executes, to elastically autoscale the allocated resources in order to maintain the desired performance even under changes in load. This rule-based mechanism relies

on the monitoring of each part of the topology to identify bottlenecks and dynamically adjusts the amount of allocated resources to comply to the user-defined performance and cost target.

There has been considerable work in the field of resource management in distributed stream processing systems, however they either emphasize on mechanisms for handling overload without obtaining additional resources on-demand, or focus more on addressing latency-related constraints. Contrarily, our work aims to provide optimal resource allocation and utilization based on high-level or low-level throughput guarantees. Moreover, our solution introduces a novel perspective to the automatic-scaling of distributed stream processing engines by providing re-usability, extensibility and faster initial decisions for the system scaling.

The contribution of this paper is summarized in the following:

- A framework for modeling the cost and the performance characteristics of streaming operators.
- A static, model-based mechanism for deciding on the right amount of resources to be allocated to each operator of the stream processing workflow according to the user-defined performance and cost constraints.
- A dynamic, rule-based mechanism for real-time auto-scaling of the allocated resources according to the monitored cost and performance.
- An open source prototype of our system ARiSto[1], which implements all the above using Apache Storm as the underlying stream processing engine.
- An experimental evaluation that proves the applicability of our methods, which are able to accurately predict the correct amount of resources required for the initial deployment of the streaming workflow and maintain the desired throughput even after sudden changes in load.

2   Preliminaries

Apache Storm [2] is the pedestal of our work. Thus, the understanding of its basic concepts and architecture [3] is of utter importance. In the following we present the basics of Storm and use the terminology introduced throughout the paper.

The core abstraction in Storm is the **stream**, an unbounded sequence of tuples that is created and processed in parallel in a distributed fashion. The tuple is a structure that can contain any kind of data from integers, byte arrays, etc. to custom objects.

A **spout** is a source of streams in a Storm application. Generally, spouts will read tuples from an external source and emit them into the Storm application. All processing in Storm applications is done in bolts. Bolts contain the processing logic and can do anything from filtering, functions, aggregations, joins, talking to databases, etc. A single bolt can perform simple stream transformations, while

---

[1] https://github.com/vgolemis/aristo.

more complex stream transformations often require multiple steps, i.e., multiple bolts. Bolts can have multiple streams as input and respectively emit more than one stream.

A **topology** is the logical graph representation of the streaming application. More precisely, it is a DAG (Directed Acyclic Graph) with spouts and bolts acting as graph vertices. Unlike a MapReduce job, which eventually finishes, a topology runs continuously, until killed. Part of defining a topology is specifying which streams feed each bolt. A **stream grouping** defines how these streams should be partitioned among the bolt's tasks.

Moving from the logical to the physical level, a subset of a topology is executed by a **worker** process, which runs in its own JVM. A running topology consists of many workers running on many machines within a Storm cluster. Each worker, tied to a specific topology, may run one or more **executors**, i.e., threads. An executor thread is spawned by a worker process and runs within the worker's JVM. Respectively, an executor may run one or more **tasks** for the same component (spout or bolt). A task performs the actual data processing. The number of tasks for a component is always the same throughout the lifetime of a topology, but the number of executors for a component can change over time. This means that the following condition holds true: #executors ≤ #tasks.

Another concept of Storm, related to the allocation of resources to each topology component is that of *parallelism* [10]. The parallelism is specified by modifying the following topology configuration parameters: (a) number of worker processes for the topology across all machines in the cluster, (b) number of executors per spout/bolt and (c) number of tasks per executor (the default is one).

All these parameters compose a configuration of the topology. Of these parameters, the number of workers and executors can be configured dynamically (i.e., after the topology is submitted). Contrarily, the number of tasks is static: additional tasks per executor do not increase the level of parallelism, since an executor always consists of one thread that it used for all of its tasks, i.e., tasks run serially on an executor. So, the only reason for having multiple tasks per executor thread is to give the flexibility to scale up the topology without taking the topology offline.

Apache Storm does not offer any mechanism to provide the optimal configuration (parallelism parameters) for user-defined topologies in an automatic manner [4]. There exist some rule-of-thumb guidelines for the parallelism of a topology, drawn from the experience of users:

- If a small number of executors per worker is defined, the cluster resources may not be fully utilized.
- If a large number of executors per worker is defined, resource contention and context switching issues may arise.
- Large number of streams create complexity in the network level.
- Performance is improved if neighbouring components are executed in the same worker because of data locality and reduced network time.
- Multiple workers per machine do not provide additional gains, only flexibility in case of worker failure.

However, these rules are quite vague and can not be used as-is for automatically defining the configuration parameters of a certain topology according to a user-defined optimization metric (e.g., throughput, latency). The user must still define the configuration herself. This can lead to a time-consuming iterative trial-and-error process in which the user runs the topology, manually monitors performance during runtime and evaluates if the configuration satisfies her expectations. Even after multiple iterations, the user may end up in a suboptimal configuration, which additionally fails to adapt to changing workload conditions due to its static nature.

3   Architecture

ARiSto is a system implemented on top of Apache Storm, that provides automatic configuration of the parallelism parameters of a topology as well as autoscaling of the provisioned resources in order to continuously comply to the user-defined throughput constraints, even after changes in load.

To this end, ARiSto employs two mechanisms: (a) a *dynamic*, on-line mechanism, that runs alongside the topology and reactively finds the optimal configuration according to the current conditions and (b) a *static*, off-line mechanism, that combines machine learning techniques with genetic algorithms to exploit knowledge from previous runs of the topology to predict the near-optimal configuration.

The common base for both mechanisms is the usage of information about the performance of each component of the topology. This information is retrieved by monitoring topology metrics. Low level component metrics (e.g., #executed tuples, #emitted tuples, latency, component capacity[2] etc.) for a given topology are periodically retrieved and converted to higher level combined metrics (e.g., execution rate, weighted average latencies, topology capacity, etc.) which can either train machine learning models or used in real-time.

The *dynamic* mechanism, given a user-defined target throughput for a topology, keeps track of its monitoring metrics and decides in real-time to adjust one or more parallelism parameters, if necessary. If it reaches the given target, it continues to periodically check to identify possible deviations from the target, due to changes in load or incoming data rate. In order to reach the given target, the mechanism needs to find the best configuration for the topology by resolving the bottlenecks. A bottleneck in a topology is a component that cannot reach the required execution rates, i.e., it cannot process the data at the rate at which they arrive from the previous processing stage. This can happen because the component uses all of its available resources and the only solution is additional parallelism to the specific component. The *dynamic* mechanism follows rules to

---

[2] The *capacity* metric, which is measured for each topology bolt and takes values between $(0,1)$, represents the percentage of time that a bolt is active (i.e., processing tuples). If this metric approaches 1 the bolt works near its maximum capacity and needs additional parallelism.

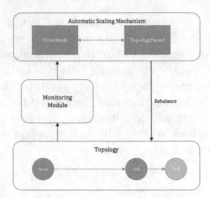

Fig. 1. The architecture of the *dynamic* component of ARiSto

decide how to modify the parallelism parameters of the topology and can be summarized as follows:

- If a component capacity exceeds a threshold th_{cc} then an executor is added for the component.
- If the number of executors per worker exceeds a threshold th_e then we increase the workers of the topology.
- If the capacity of individual components is normal but the total capacity of the topology exceeds a threshold th_{high} then we increase the workers of the topology.
- If the capacity of individual components is normal and the total capacity of the topology is below a threshold th_{low} then we decrease the workers of the topology.

After any change performed in the topology's parallelism parameters we grant a time window for the topology to stabilize (warm up phase). During this window we observe the throughput of the topology and proceed with further changes if required.

The architecture of the *dynamic* module of ARiSto is depicted in Fig. 1. The *TopologyParser* manages the high-level functionalities like invoking the *Monitoring Module*, reading the user configuration and initializing the data structures. After the initialization it is responsible for calling the *FlowCheck* periodically. The *Flowcheck* manages the low-level functionality of the mechanism like checking if the target is reached and changing the parallelism parameters if needed.

The *static* mechanism can be given user level constraints (money, time, etc.) and finds the near-optimal configuration(s) that comply to the user-defined constraints. To be able to predict the best configuration for any given topology the mechanism should be able to predict the performance for every possible configuration. To this end it relies on machine learning and heuristic techniques. The basis of the particular mechanism is machine learning models for the components of the topology. In order to create the models for every component of a topology, each component goes through offline profiling which aims to train the

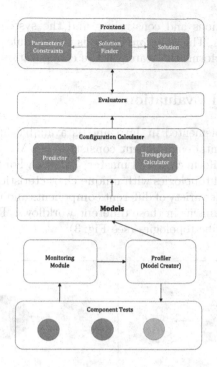

Fig. 2. The architecture of the *static* components of ARiSto

models. More extensive and focused training leads to more accurate models. The choice to create the models at the component level and not the topology level was made to render the mechanism flexible and expandable: Multiple topologies may use the same components, thus, the component models can be used by multiple topologies and enriched. In that manner, we create libraries of component models which are reusable and extendible. After the required models are created, they are used to predict the performance for any given configuration of a topology containing the specific components. In particular, a prediction is given by each model for its corresponding component and all of them are combined to produce the final performance prediction for the configuration of the topology. Subsequently, a heuristic algorithm based on genetic algorithms (NSGA-II in our case) that searches the space of all possible configurations is executed to find the near-optimal one.

Figure 2 presents the architecture of the *static* mechanism of ARiSto. The *Models* are created per component (using the WEKA [12] framework), initially through offline profiling by the *Profiler*. The *Predictor* relies on the *Models* to predict the throughput of individual topology components. The *Throughput Calculator* module calculates the total topology throughput by combining the predictions of individual components, given a specific topology configuration.

The *Evaluators* are those who build the high level constrained problem based on the user level constraints. The *Front-end* is the API we expose. The user defines her configurations and constraints and the system returns the list of optimal configurations. The *SolutionFinder* is the module that creates the correct evaluator and performs the search in the configuration space.

4 Experimental Evaluation

ARiSto has been implemented in Java and is available under an open-source license. The experimental environment consists of 8 VMs (2 cores/4 GB RAM) in order to create a uniform cluster (1 master, 7 slaves). For the evaluation of our system we assembled 4 topologies with unique characteristics and structure. Our target is to cover a wide variety of different components and flows and observe the behavior of our mechanisms in these different workflows. These include custom made as well as scientific topologies (see Fig. 3):

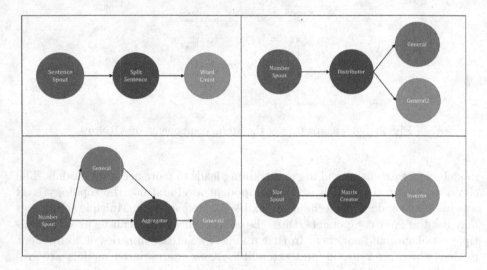

Fig. 3. The topologies of the experimental evaluation

- **WordCount**, one of the fundamental examples in distributed environments. (CPU intensive, Network intensive workflow)
- **CyberShake**, a scientific workflow used by the Southern California Earthquake Center to characterize earthquake hazards in a region. (Network intensive workflow)
- **Montage**, a scientific workflow created by NASA/IPAC stitches together multiple input images to create custom mosaics of the sky. (CPU intensive, Network intensive, Memory intensive workflow)
- **Matrix**, a custom-made workflow that performs matrix operations. It is customizable and easily expandable. (CPU intensive, memory intensive workflow)

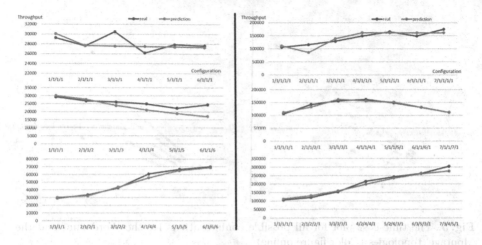

Fig. 4. Evaluating the total topology throughput when varying the #workers, #executors and both (vertically) for the Wordcount and the Montage Topologies (horizontally)

We evaluate ARiSto in terms of sensitivity to the parallelism-related configuration parameters, efficacy of the *static* mechanism and the accuracy of its predictor module.

Performance impact when varying the parallelism parameters. In the first set of experiments (first row of Fig. 4), we gradually increase the workers without changing anything else in the configuration. For the WordCount Topology we can see that by adding a worker to the initial configuration we lose in terms of performance. This happens because we lose the data locality (if all the components are executed in the same worker we avoid over-the-network data transfer). From that point on, additional workers do not improve the performance because it is bound by the main performance bottleneck i.e., the SplitSentence bolt. For the Montage Topology, as we provide more workers we can see a slight performance increase. This is due to the fact that the workflow is more cpu- than network-intensive.

In the second set (second row of Fig. 4), we gradually increase the number of executors for one of the components alongside the worker increase. For the WordCount Topology we increase the WordCount component executors which is not the bottleneck component. We can see that the topology performance immediately deteriorates. This happens because we increase the executors of a component which was not overloaded and we thus create additional network complexity for the SplitSentence component, which is the actual bottleneck. For the Montage Topology we increase the Aggregator component executors, which is one of the initial bottlenecks. We can see a gradual improvement in the topology performance but after several steps the performance starts to fall. This happens because we initially solved the cpu bottleneck (Aggregator), but by adding more executors of the Aggregator component we created a network bottleneck in the previous processing stage.

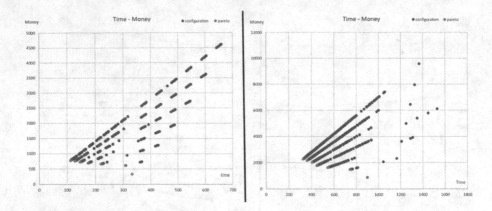

Fig. 5. The time-cost plot for all possible configurations for the Wordcount and the Montage Topologies (Color figure online)

In the third set (third row of Fig. 4), we gradually move to better configurations that simultaneously solve all the major bottlenecks that the topology is facing. For the WordCount we add executors for the SplitSentence component, which solve the cpu bottleneck and at the same time we add executors for the WordCount component, which improve the data locality and network complexity. For the Montage Topology we add executors for the Aggregator component, which solve the cpu bottleneck and at the same time we add executors for the General1 component which solve the network bottleneck and provides better data locality.

Efficacy of the static mechanism. For the 2 topologies presented above, all possible configurations are plotted in a time-cost plot (time and cost are two of the most important user constraints) in Fig. 5:

The orange dots are the configurations selected by ARiSto and they coincide with the pareto optimal configurations. Pareto optimal is a configuration that offers the best trade-off between time and cost.

The static mechanism is also evaluated based on its prediction accuracy. The accuracy is calculated as the difference between the predicted and actual performance for a given topology and configuration. The accuracy evaluations of Fig. 6 concern the CyberShake and Montage Topologies respectively. We can see that the mechanism has a discrepancy below 10% in approximately 60% of its predictions and below 20% in approximately 90% of its predictions.

The desirable behavior of the mechanism is to be very accurate in regions of global performance maxima. The regions of local/global minima as well as the majority of local maxima are not of such importance. The predictor we implemented generates very accurate predictions for the global maxima and the transitional regions, is less accurate for local maxima and purposefully underestimates the global minima and part of the local minima. This is the reason that we have a percentage of prediction with high discrepancy. The decision to underestimate

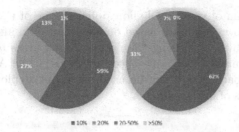

Fig. 6. The accuracy of the static mechanism for the CyberShake and Montage topologies

the local/global minima was made in order to have a faster search in the configuration space by the SolutionFinder module. When it searches a region that is dramatically below the performance the heuristic algorithm will not search again in the particular region. Additionally, solid and accurate transitional regions will help the algorithm find the local and global maxima.

5   Related Work

Distributed stream processing has been an active area of research over the last decade, when the need for real-time analytics over vast amounts of data became more prominent and called for scalable streaming engines that could handle them. Several such engines emerged, either proprietary such as S4 [8], MillWheel [13] and DataTorrent [5] or open source, such as Storm [2], Heron [17], Spark Streaming [9] and Flink [1]. None of these engined inherently provide auto-scaling capabilities to meet user-defined performance constraints with minimum amount of provisioned resources.

Works related to resource management in distributed stream processing systems mostly emphasize on mechanisms for handling overload: load-shedding [20], admission control [21], adaptive query planning [19], load balancing [22] and efficient initial operator placement [16] do not address overload by obtaining additional resources on-demand.

The concept of elastic scaling of distributed stream processing systems has been studied in recent works. However, they focus on the latency aspect of the system following different directions like optimizing operator movement during scale-in/out of the system to eliminate latency spikes [15] or using reactive strategies to enforce latency guarantees in scalable stream processing systems [18]. ARiSto aims to provide optimal resource allocation and utilization based on throughput guarantees. Moreover, our solution follows a novel, model-based approach which provides reusability, extensibility and faster initial decisions for the configuration of the resource-related parameters.

The work closest to ours is the very recently published Dhalion [14], a system deployed on top of Heron [17], which provides with self-regulating capabilities through the execution of various Dhalion policies. One of them is to automatically scale up and down resources based on the input data and another to auto-tunes the topology by provisioning the necessary resources to meet a throughput SLO. Dhalion employs only rule-based methods, while ARiSto also follows a model-based approach, which can be used when first deploying a topology to speed up the process of auto-scaling.

6   Conclusions

In this paper we presented ARiSto, a system that provides automatic scaling of resources in Apache Storm in order to comply to the user-defined performance and cost constraints, by virtue of two mechanisms, a static and a dynamic one. The static mechanism relies on performance an cost models of the streaming operators that are involved in the workflow topology and based on genetic algorithms it decides on the exact amount of resources to be allocated to each one of them. The dynamic mechanism is able to adjust to current load conditions, maintaining the desired throughput/latency with minimum cost (i.e., resources). This is achieved by monitoring all topology entities, identifying the bottlenecks and adding/subtracting resources according to a set of rules. The evaluation of our prototype implementation showcases the ability of our system to efficiently decide on the resource-related configuration parameters, maintaining the desired throughput at all times.

References

1. Apache Flink. https://flink.apache.org/
2. Apache Storm. http://storm.apache.org/
3. Apache Storm Concepts. http://storm.apache.org/releases/1.0.0/Concepts.html
4. Auto-Scaling Resources in a Topology. https://issues.apache.org/jira/browse/STORM-594
5. DataTorrent. https://www.datatorrent.com/
6. Healthcare Looks to Real-Time Big Data Analytics for Insights. https://healthitanalytics.com/news/healthcare-looks-to-real-time-big-data-analytics-for-insights
7. Real-Time Stream Processing as Game Changer in a Big Data World with Hadoop and Data Warehouse. https://www.infoq.com/articles/stream-processing-hadoop
8. S4 Distributed Data Platform. http://incubator.apache.org/s4
9. Spark Streaming. https://spark.apache.org/streaming/
10. Understanding the Parallelism of a Storm Topology. storm.apache.org/releases/1.0.0/Understanding-the-parallelism-of-a-Storm-topology.html
11. Use Cases for Real Time Stream Processing Systems. https://dzone.com/articles/need-for-using-real-time-stream-processing-systems
12. Weka 3: Data Mining Software in Java. http://www.cs.waikato.ac.nz/ml/weka/
13. Akidau, T., et al.: Millwheel: fault-tolerant stream processing at internet scale. In: VLDB 2014, vol. 6, no. 11, pp. 1033–1044 (2013)
14. Floratou, A., et al.: Dhalion: self-regulating stream processing in heron. In: VLDB 2017 (2017)

15. Heinze, T., et al.: Latency-aware elastic scaling for distributed data stream processing systems. In: DEBS 2014, pp. 13–22. ACM (2014)
16. Kalyvianaki, E., Wiesemann, W., Vu, Q.H., Kuhn, D., Pietzuch, P.: SQPR: stream query planning with reuse. In: ICDE 2011, pp. 840–851. IEEE (2011)
17. Kulkarni, S., et al.: Twitter heron: stream processing at scale. In: SIGMOD 2015, pp. 239–250. ACM (2015)
18. Lohrmann, B., Janacik, P., Kao, O.: Elastic stream processing with latency guarantees. In: ICDCS 2015, pp. 399–410. IEEE (2015)
19. Markl, V., et al.: Robust query processing through progressive optimization. In: SIGMOD 2004, pp. 659–670. ACM (2004)
20. Tatbul, N., et al.: Load shedding in a data stream manager. In: VLDB 2003, pp. 309–320. VLDB Endowment (2003)
21. Wolf, J., Bansal, N., Hildrum, K., Parekh, S., Rajan, D., Wagle, R., Wu, K.-L., Fleischer, L.: SODA: an optimizing scheduler for large-scale stream-based distributed computer systems. In: Issarny, V., Schantz, R. (eds.) Middleware 2008. LNCS, vol. 5346, pp. 306–325. Springer, Heidelberg (2008). https://doi.org/10.1007/978-3-540-89856-6_16
22. Xing, Y., Zdonik, S., Hwang, J.-H.: Dynamic load distribution in the borealis stream processor. In: ICDE 2005, pp. 791–802. IEEE (2005)

Author Index

Printed in the United States
By Bookmasters